Contributions to Economics

For further volumes:
http://www.springer.com/series/1262

Giorgio Calcagnini • Ilario Favaretto
Editors

Small Businesses in the Aftermath of the Crisis

International Analyses and Policies

Physica-Verlag
A Springer Company

Editors
Giorgio Calcagnini
Università di Urbino "Carlo Bo"
Urbino
Italy

Ilario Favaretto
Università di Urbino "Carlo Bo"
Urbino
Italy

ISSN 1431-1933
ISBN 978-3-7908-2851-1 ISBN 978-3-7908-2852-8 (eBook)
DOI 10.1007/978-3-7908-2852-8
Springer Heidelberg New York Dordrecht London

Library of Congress Control Number: 2012942853

Printed on acid-free paper

Physica-Verlag is a brand of Springer
Springer is part of Springer-Science+Business Media (www.springer.com)

Preface

SMEs are the backbone of the Italian entrepreneurial system and have provided the main impetus for economic development in the past decades. Notwithstanding their structural weaknesses, SMEs remain the platform on which the Italian economy should build new growth processes. Some SMEs should necessarily undertake strategies to increase their size and their degree of internationalization, to become the driving force behind the Italian production system; others should exploit their own areas of expertise within intelligent production chains that will be able to inherit what still remains of the industrial district logic on which the Italian economic development was based before the globalization process began.

It is imperative that all SMEs choose to re-position themselves in competitive terms, making up for time lost over the last decade. The price of doing otherwise would be exclusion from all markets.

Supporting SMEs on their difficult path towards rebuilding is in the interest of the entire nation, and the role of banks in this context remains fundamental. Banks are responsible for assuring adequate credit flows towards the firms that will be able to generate a new phase of investments as soon as they are capable; this is necessary to foster profound innovation in our production structure. Investments aimed at incorporating growing amounts of new technologies in the production and management processes of SMEs are needed to fill the obvious productivity gap in their production factors with respect to larger firms. But also investments to increase what we commonly call immaterial or intangible capital. Factors such as human capital, social capital, the propensity to form networks, the ability to express one's qualities and to adapt products to demand, etc., are not well managed by SMEs in terms of key strategy. However, they appear to be fundamental elements so that the small size of these firms does not remain an insurmountable obstacle to their internationalization prospects.

The activities of conducting research and facilitating international dialogue which the University of Urbino has for some time – through the commitment and passion of Professors Calcagnini and Favaretto – dedicated to topics connected with SMEs thus represents a contribution of the highest order for entrepreneurs, their professional organizations, policy makers, and the banking system, in that it provides a growing and wide spectrum of elements on which to reflect and instruments with which to operate. The call that rises from the various papers that make up this book concerns the urgency with which our country should necessarily, – without wasting time – make those entrepreneurial decisions and create those legal and infrastructural requirements to regenerate the system of small- and medium-sized firms. The above-mentioned changes are necessary so that SMEs can compete and create adequate employment levels, thus making an incisive contribution to the growth of our economy.

Luciano Goffi

Director-General of the
Banca Popolare di
Ancona, Jesi

Introduction

The economic crisis, though dramatic for many people, has represented an incredible opportunity for scholars studying different aspects of how modern economies function. It has been a sort of "ideal" randomized experiment by which testing business models has further developed during recent years.

This book contains papers presented at the third international conference on small- and medium-sized businesses (SB) held at the Università di Urbino "Carlo Bo" in October 2010. At that time we were excessively optimistic, given that the conference theme was how SBs responded to the crisis (and the state in which they found themselves in the aftermath of the crisis), thus assuming that the latter was mainly over by then. If this were the case, we also expected to discuss new production and business models. Unfortunately, as of this writing, the crisis is still affecting the world economy. Many businesses found closing down a profitable option, while entrepreneurs who are still in business are working hard to adapt their companies to the new economic context.

Economies have different business structures around the world, and it is not our purpose here to analyze them in turn. We will again refer to the Italian economy as a laboratory, given the large number of small-sized businesses that characterizes it. The task of discovering differences across countries in the analyses contained in the chapters of this book is left to the reader.

One of the results that emerged from the two previous conferences (see Calcagnini and Favaretto, 2011) was that economies characterized by the presence of small-sized firms are more vulnerable to shocks than countries where the average size of businesses is larger. There may be two main explanations behind the relationship between firm size and GDP growth rates. In the case of Italy, they are strongly connected and concern, on one hand, the structure of the production system and, on the other, the competitive position of Italian firms in international markets.

Notwithstanding the restructuring carried out by Italian manufacturing firms during the period 2003–2007,[1] at the outbreak of the current economic crisis the

[1]See de Nardis-Pappalardo (2010).

Italian economy (industry) still found itself with a production system highly skewed towards small- and very small-sized businesses. Being dependent upon small sized firms has two implications: (a) a large share of firms make their production decisions based on domestic demand; (b) small-sized firms find competing internationally more difficult than larger firms. The combined effect of a compressed domestic demand and the increased competition in international markets may explain trends in Italian GDP growth rates during the crisis. Moreover, by expanding the time horizon to a longer run scenario, this hypothesis can describe the entire period from the end of the 1990s onwards.[2]

Further, the economic crisis and its persistence caused the smallest and least efficient businesses to exit the market. Simultaneously, many production chains dissolved, causing bankruptcies among sub-suppliers. The latter exited the market because they lost most of their customers and not because they were inefficient. In other words, the crisis negatively affected the labor division among most of Italian manufacturing firms, even though its effects had diversified intensities depending on the industry type, as well as the country in question.

Firms that positively reacted to the economic recession, and were still able to compete internationally followed two strategies. On one hand, as an answer to the dissolving of production chains, firms reorganized their production cycle by re-internalizing some of the phases that had been previously outsourced or decentralized. Consequently, the whole business organization has been restructured in the attempt to reach a higher level of efficiency in capacity utilization and economies of scale. On the other, firms have been starting a new wave of product and process innovations that is positively affecting their cost structure. This strategy was implemented by investing in intangible assets and also by hiring more skilled workers, even though mainly diffused among medium- and large-sized firms. Both organizational and product innovations pay witness to the fact that the current reduction in the number of very small-sized firms still characterizing the Italian manufacturing industry is coherent with its need to further consolidate firms. Therefore, it is likely that small-sized firms will play a less important role within a modern and competitive system conditioned by global markets. This is so even though all subjects involved in market labor division (i.e., buyers and suppliers) should cooperate to favor firms reaching higher equity levels, investing larger amounts of resources in human capital, and developing a more modern finance culture among their managers and consultants. If the former changes were made without a strongly coordinated and coherent industrial policy at the central and local levels, competing domestically and internationally will be increasingly difficult for Italian manufacturing businesses.

This brings us to the second explanation, still related to business size, which focuses on the internationalization process of the Italian economy. By internationalization we mean trends in exports, imports, and foreign direct investment (FDI). It has been shown that internationalization is positively correlated to the number of

---

[2]See also Rey-Varaldo (2011).

firms actually involved in international activities (i.e., the 'extensive margin'). This 'extensive margin' is much more important than the 'intensive margin', that is the average exports, imports, and FDI per firm. Therefore, successful internationalization is much more about increasing the number of firms involved than about, for example, increasing the involvement of firms already exporting. Policy oriented to increase the 'extensive margin' should, therefore, improve firm performance in terms of employment and productivity.[3] Further, firms that are larger in size have a more-skilled workforce, and are more innovative and productive. Hence, they are more likely to export than others. This pattern about the role of firm characteristics on the 'extensive margin' is very similar across countries. Therefore, different industrial structures explain differences in countries' international performances since similar firms behave similarly across countries.[4]

Recently, the Italian National Institute of Statistics (ISTAT) published new national accounts for the period 2000–2010. Changes between the old and new series occurred because of the adoption of the NACE-2007 industry classification and the CPA 2008 product classification associated with the industry classification. In the case of exports and imports ISTAT calculated price indexes instead of the traditional unit, i.e. average value indexes.[5] The use of the new foreign trade price indexes to calculate exported and imported quantities shed new light on Italian external trade, especially on exports as an indicator of industry competitiveness.

The consolidated view according to which, during the 2000s, Italian industry experienced a decline with respect to the previous decades and other European countries does not find support in export growth rates (see Table 1). The latter were only negative and quantitatively larger than those of the other economies shown in Table 1 during the most recent years of the economic crisis. However, considering the whole period, the cumulated export growth rate is the same size as the French one. Further, when the 2003–2007 period is analyzed, Italian export growth was significantly larger than that of France and Spain, but 40% lower than Germany's (see Table 1). These were the years during which Italian industry underwent a restructuring that re-allocated resources from less to more efficient firms, from traditional consumption goods industries to ones with investment and intermediate goods. This resource re-allocation involved industries showing both comparative advantages and disadvantages.[6]

In summary, Italian industry shows that at its core there is a significant group of firms that are trying to react to changes in their competitive environment and to the worst economic crisis since 1929. They aim to compensate for stagnant domestic demand by looking for new foreign markets or by expanding their existing ones. However, the quest for growing foreign markets is a complex one because of the relatively small size of Italian businesses. It would be ungenerous not to recognize

---

[3]See Mayer-Ottaviano (2007), pp. 4–5.

[4]See Barba Navaretti et al. (2010), pp. 4 and 9.

[5]See ISTAT (2010).

[6]See de Nardis-Pappalardo (2010), p. 4.

Table 1 Cumulated export growth rates (quantities)

	2000–2010	2000–2007	2003–2007	2007–2010
France	14.52	17.99	15.09	−3.47
Germany	55.37	52.60	39.43	2.78
Italy	14.77	22.87	24.35	−8.10
Spain	27.54	29.94	20.13	−2.40

Source: our calculations based on ISTAT (Italy) and IMF (France, Germany, Spain) data

the effort made by those small firms that have internationalized, even though there are still too few of them. Further, in most cases, internationalized small-sized firms face the challenge of foreign markets by means of traditional trade strategies and long distribution chains that provide thin profit margins. Indeed, Italian manufacturing firms are traditionally successful in production, but less in marketing, in distribution chains, and in controlling final market outlets. Therefore, they are able to retain only a small share of the whole product market value they generate.[7]

The challenge taken up by Italian industry needs to see a redoubling in firms' efforts towards a more intensive use of scientific and technological knowledge, plus more skilled workers as well as increased equity among their financial resources. This process should necessarily go along with an increase in firm size that does not mean that firms should become large-sized. Firms with around 50 to 100 employees (medium-sized firms) contribute greatly to global exports and make up the backbone of export performance for most European countries.[8] However, even among firms with 20 to 50 employees more than 60% are exporters. This means that even among small-sized firms productivity is high enough to compensate for export costs.[9]

These results strongly support conclusions we reached in Calcagnini – Favaretto (2011) where we sketched a few simple policy guidelines in favor of the consolidation of Italian firms. The latter will bring about an increase in firm size that, as shown above, is positively associated with higher levels of R&D investment, innovation, skilled employment, more effective foreign trade and FDI policies, and the opportunity to access financial resources coherently with all the decisions typical of modern companies. These are a few issues that should be the primary concern of a modern industrial policy oriented to favoring the competiveness of Italian firms. This without speaking of other regulation reforms focused on creating a more favorable environment for "doing business".[10]

The contributions included in this book are ideally divided in two groups. The first group focuses on the effects of the economic crisis on the 'real side' of small

---

[7]See Rey-Varaldo (2011), p. 750.

[8]See Barba Navaretti et al. (2010), p. 9.

[9]See de Nardis-Pappalardo (2010), p. 28.

[10]To understand the negative role played by Italian regulations on the start-up and development of firms, see World Bank (2011).

business economics, such as innovation, technologies, job creation, internationalization, economic policies. The second one contains analyses concerning the 'financial side' such as access to and the cost of credit and the existence and the role of 'internal' capital markets.

Arzeni, Cusmano and Potter discuss the policy environment that has emerged through the recent crisis and the policy challenges ahead, moving from the impact of the crisis on SMEs and entrepreneurs across OECD and non-OECD countries. They highlight the fact that SMEs and entrepreneurship can greatly contribute to the general economic recovery, and discuss key policy areas for long-term growth, as identified at the international level by governments, SMEs and stakeholders through the OECD Bologna Process on SME and Entrepreneurship Policies. The authors assert that fostering entrepreneurship and small business development are key ingredients in any development strategy. The recent crisis and its effects on the SME sector and potential entrepreneurs have, if possible, further highlighted the crucial role of small businesses and start-ups in the creation or preservation of employment, as well as in the generation and diffusion of innovation. New and young firms can translate knowledge and ideas into jobs and wealth, often exploiting opportunities that have been neglected by more established companies. However, for entrepreneurial dynamics to sustain innovation, employment and economic expansion, the survival and growth of new firms is crucial. In other words, what matters for growth is not only the entry of new entrepreneurs into markets, but also the quality, type, potential and sustainability of the entrepreneurial projects launched. Therefore, Arzeni et al. highlight some of the most important policy issues and measures that governments should adopt to unleash entrepreneurship and sustain competitive SME development. Among these short- and long-term measures they stress the need for drastic reduction in the length of time the public administration and large firms take to pay small suppliers of goods and services, and the increased simplification of administrative procedures required of small businesses. Further, and most important, would be the necessity to support the creation of SME networks to facilitate their access to public procurement, their international growth, and to upgrade their human capital skills to capture the intellectual assets that are vital for their growth.

Zecchini describes the OECD countries' policy approaches to the SME crisis during the recent economic recession. After identifying the causes of SME weakness and the channels through which the recession affected them, the government policy approach is spelled out and its impact is evaluated, leading to a number of conclusions. The main structural sources of SMEs' greater vulnerability are: (1) their relatively higher sensitivity to variability in sales; (2) their lower market power by comparison with large firms; (3), their relatively lower propensity to engage in R&D and innovation projects; (4) their relatively weaker capital structure. Of course, SMEs are a highly diverse group of firms. Among SMEs overall, small-sized firms and micro firms are the most vulnerable. Medium-sized firms are more capable of entering foreign markets and investing in R&D. Innovative small firms can grow very rapidly and defend their market shares thanks to intellectual property rights. Zecchini believes that the experience to date with government polices for SMEs in response to the recession leaves a number of issues still requiring a durable solution: (1) ways and means to make SMEs less vulnerable to changing conditions in bank lending; (2) the relationship between banks and SMEs tends to be ruled by the automatic application of backward-looking scoring and rating methods to assess credit merit; (3) when devising assistance programs, small-sized firms should be treated separately from medium-sized firms, because their needs differ significantly; (4) small-sized firms can grow best if they act in a functional relationship with other firms, particularly larger companies; (5) a key to strengthening a firm's competitive position during and after a recession lies with research and innovation; (6) regarding international trade, Government support to SMEs during the recession consisted chiefly in expanding export credit and credit guarantee facilities.

The Basalisco and Rey paper introduces the 'platform industrial policy' concept, that is a policy equivalent of the business model construct and considers the application of a platform strategy to promote the adoption of advanced network infrastructures by small- and medium-sized enterprises (SMEs). Indeed, in countries where a multitude of SMEs compose the backbone of the economy, the presence of free riding and counterpart uncertainty makes the coordination of Information and Communication Technology (ICT) of investment extremely complex and obstructs the network adoption dynamics. Therefore, the authors analyse – with specific reference to the Italian case – the incentives of industrial players in a context where SMEs perform an important economic role and where traditional, rather than high-tech industries, are predominant. Further, they review the contributions provided by the economic and managerial literature on platforms or two-sided markets. They leverage this by proposing a policy application of the platform concept and show its potential benefits in allowing the manufacturing sector to gain from access to the service economy. They classify platforms according to their function and identify those types that can serve the purpose of industrial promotion, each of which stresses the importance of their role. Basalisco and Rey identify three platform types: (a) Platform A – Economics perspective that should help managing transactions across different sides of the market that are linked by cross-market externalities and platform prices; (b) Perspective B – The platform as a standard that should help enable the creation of new services and markets by promoting the adoption of a core platform technology and coordinating technological development and the necessary standards; (c) Platform C – The platform as an open supply chain that should help open the supply chain of a set of key (or pivotol) firms by enabling a greater degree of independent interaction between these firms' suppliers (e.g. SMEs) and a broader set of product end users.

This platform classification leads them to review aspects of business governance that are paramount to successful platform functioning, given the policy context. The authors are interested in describing how traditional industrial districts should develop into business districts where the distribution of benefits and risks among the various components takes into account their contribution to the creation of the business value added. Therefore, in their opinion, platform governance should provide the means to address this distribution issue by the adoption of common rules.

Platform governance concerns the exercise of market power, but also involves a political consideration, that is, a shared understanding within the community of platform stakeholders of the division of power and responsibilities concerning the strategy and conduct of the platform itself.

Reedy and Strom analyze America's Slow Leak in Job Creation. Although understandable in light of its traumatic impact, the Great Recession of 2007–2009 may be distracting attention from a more fundamental troubling economic trend. The United States appears to be suffering from a long-term leak in job creation that preceded the 2007–2009 recession and has the potential to persist for an unknown length of time. The heart of the problem, according to the authors, is a pullback by newly created businesses, the economy's most critical source of job creation, which are generating substantially fewer jobs than one would expect based on past experience. Two implicit assumptions in the debate about jobs, which they believe to be wrong, deserve mention. First, policymakers' focus on big changes in employment because of events such as a new manufacturing plant or the recruitment of a business to a community ignore the more important fact that the jobs outlook will be driven more by the collective decisions of the millions of young and small businesses whose changing employment patterns are not as easy to see or influence. Second, it is just as easy to be deluded into thinking that the jobs problem will be solved by growth in the number of the self-employed. The last decade has brought about some fundamental shifts in U.S. labour markets, including the rise of outsourcing, and not just to foreign locations. In many cases, companies or individuals that once would have been hired as employees of a business now are performing the work on a temporary basis as contractors through other professional service organizations or under their own self-employment contracts. These individuals, while sometimes characterized as entrepreneurs, are not likely to employ others or to reach significant scale. Therefore the clear challenge for the U.S. economy is to start more employer businesses, ensure that they are starting larger, and nurture their growth.

Calcagnini and Favaretto analyze trends in the Italian production structure before and during the current economic crisis, and changes in the decentralization process following the most recent technological innovations and functional reorganization among firms. More specifically, they describe how the latter has been slowly assigning an increasing role to better-structured firms with respect to the past. The authors show that the long-time quest for the firm size most suited to the Italian economy has recently taken a radical change of direction. The new competitive context favors those firms that more intensively invested in strategic variables such as innovative technologies, human capital, and complex organizations within which each individual firm coordinates and guarantees the production phases characterized by higher quality content. As a consequence, the authors describe the transformation occurring in the Italian production system from a situation characterized by a very large number of small- and very small-sized firms, low innovation levels, and a large share of irregular economy into a new system where firms progressively find investing in R&D, participating in inter-industry organizations, selecting more educated employees and providing better training to be rewarding. Furthermore, just as important would be for firms to increase their size to become more profitable. All of the aforesaid would allow firms to cope with stricter, more competitive markets at the international level. This transformation is necessary to allow Italian firms to keep competing internationally, even though it will likely come at the cost of a smaller production base (a lower number of firms). The changes outlined by Calcagnini and Favaretto are be in line with EU policies supporting regional and endogenous development projects that make the most of the above-mentioned strategic variables.

Cucculelli uses longitudinal data from a sample of 142 Italian family firms to examine how a firm's decisions of strategic relevance – like the introduction of a new product and the succession process after a company's founder steps down – are affected by short-term external factors and firm-specific factors. The empirical analysis exploits the advantages of a conditional risk set duration model, which allows taking into account the multiple ordered occurrences of similar failures (multiple product introductions and successions of managing directors) during the life of a company. The empirical analysis shows that economic downturns affect firms' decisions substantially, as they change the timing of product introduction and succession or stop them from being adopted at all. The probability that both events will occur increases steadily as the negative effects of downturns vanish, but at different rates: delays in succession are shorter than those in product adoption. Finally, succession only induces new product introduction when the first happens in favourable years, whereas new product introductions always reduce the probability of having a succession in management. However, this negative impact is less evident in positive years, given that the pro-cyclical nature of product introduction has become more evident.

Rombaldoni investigates the processes of internationalization followed by Italian firms as a path towards sustainable growth after the recent economic and financial crisis. However, the facets of such a process are numerous, complex and often closely related: delocalization, fragmentation of the productive process, technology and innovation transfers and the globalization of production chains, are just some of the most significant aspects. Production delocalization is one of the most frequent answers to the growing pressures of globalization: if many micro-enterprises have been destroyed by the crisis, for many others the process of qualitative upgrading of products, as well as the strategic process, are proceeding. The most dynamic growth of outgoing productive investments has to be ascribed to medium-sized companies, thus confirming the relationship between size and the complex organization of activities. As far as internationalization and innovation are concerned, empirical evidence for Italy highlights that the way in which a business is present in international markets makes a difference: firms resorting to cooperation agreements are able to generate more innovation compared to their colleagues localized in national markets, and this is due in part to massive knowledge input, but also to the former having greater access to external ideas. The right investment mix in innovation, organizational and technological improvements, as well as participation in a network and an active role in the global value chain, are essential so that internationalization strategy becomes an opportunity and not a threat.

The Gramigna article provides general background on the events that lead to the financial crisis and Great Recession of 2008–2009 in the United States. A strong argument can be made that the origins of the crisis can be traced to a fundamental imbalance between house prices and personal income. This imbalance resulted in a pronounced drop in inter-bank confidence, leading to a dramatic rise in the relative cost of capital for financial institutions, and a sharp reduction in lending. As a result, the economy experienced a pronounced recession, hitting small-sized businesses particularly hard. The U.S. federal government responded with enormous conventional and non-conventional measures to assist the financial markets and stimulate the economy. In addition, the federal government undertook numerous actions to address the sharp drop in the small business sector by providing additional incentives to stimulate small business lending and fortify small businesses' balance sheets. The U.S. Small Business Administration (SBA) played a critical role in increasing access to capital for small businesses. The SBA's Capital Access Programs experienced an approximate 60 percent decline in its lending volume. The SBA addressed this challenge by providing additional incentives to financial institutions and by assisting in the unfreezing of the secondary market for SBA loans.

Ferrando analyses the access to finance enjoyed by SMEs in the euro area and investigates the issue of whether SMEs have suffered more than large-sized firms during the crisis. The author uses information directly provided by the firms through a new firm-level survey based on a sample of non-financial corporations in the EU (SAFE survey). The findings seem to point out that banks followed a more cautious lending policy towards younger and, hence, smaller firms, possibly because of their shorter track record and reputation. Since SMEs are often unable to switch from bank credit to other sources of finance, experiencing major financing obstacles can be a considerable challenge for them with respect to larger firms. The results of a more analytical exercise that investigates the underlying determinants of financing obstacles show that while the general perception of heightened financial obstacles was broadly based across firms during the recent crisis, age and ownership structure are important explanatory variables for firms' perceived financing obstacles across countries. As there is a strong relationship between age, simple structures of ownership and the size of firms, the results confirm the fact that SMEs were indeed hit harder when banks' credit standards tightened.

Mach and Wolken's contribution examines the effects of credit availability on small-sized firm survivability over the period 2004–2008 for non-publicly traded small-sized enterprises. Using data from the 2003 Survey of Small Business Finances, they develop failure prediction models for a sample of small firms that were confirmed to have been in business as of December 2003, with particular attention to the impact of credit constraints. They find that credit constrained firms were significantly more likely to go out of business than non-constrained firms. Moreover, credit constraint and credit access variables appear to be among the most important factors predicting which small U.S. firms went out of business during 2004–2008. Other significant characteristics include location—i.e. being headquartered in the West North Central, East North Central or South Atlantic Census divisions—and industry—i.e. being a construction or retail trade business. Equally informative is the finding that generally financial ratios and geographic controls were not significant in predicting whether the firm would still be in business in 2008. The authors think that the recent economic turmoil is not representative of all periods: despite the extraordinary times, credit access measures do seem to be predictive of firm performance.

Iacobucci analyzes whether small- and medium-sized firms belonging to groups are more or less financially constrained in their investment policy than their stand-alone counterparts, and whether there are differences in the investment-cash flow sensitivity between affiliated and non-affiliated firms. Business groups are preferred to other organizational forms because the legal autonomy of member companies allows entrepreneurs to implement the desired mix of control and autonomy in developing and managing a portfolio of businesses. In small- and medium-sized groups, those businesses are often closely related to one another, thus allowing the controlling entrepreneur to exploit technological and marketing synergies between controlled companies. Managing the cash flow in a unitary way generates one of the ways in which those synergies can be exploited within the group, thus creating an internal capital market. This means the possibility of tunneling financial resources from cash-producing to cash-absorbing companies, thus smoothing the impact of external shocks and taking advantage of firms that are in different stages in their business development. The internal capital market also offers the advantage of reducing the need to raise funds from external sources in order to finance new investment. The empirical part of the paper, based on the Unicredit dataset, tests the hypothesis of the presence of an internal capital market in small- and medium-sized groups by comparing the cash-flow investment sensitivity between affiliated and non-affiliated companies. Overall the results confirm that firms belonging to groups are less dependent on the internal generation of funds when financing their investments. To the extent that cash-flow investment sensitivity is interpreted as the presence of financial constraints, belonging to a business group reduces the constraints when raising funds to finance investments in tangible assets.

Calcagnini, Farabullini and Giombini study the role of guarantees on interest rates charged on producer households' loans by analyzing individual Italian bank and producer household data drawn from the Central Credit Register at the Bank of Italy over the period 2006–2009. Further, the paper tries to understand bank behavior before and during the recent financial crisis.

As informational opacity is expected to be negatively correlated with firm size, small-sized firms are more often affected by either credit rationing and/or a higher interest rates than larger firms. These two phenomena are typically exacerbated during financial crises. During the period 2008–2009, the number of loans to Italian small businesses decreased by more than 30%, as opposed to a 20% decrease in the case of all firms (firms and producer households).

Firstly, the paper estimates a bivariate probit model to capture how the crisis, together with the observed riskiness of borrowers and other variables influenced the probability of observing secured loans. This analysis shows that the probability that loans would be secured increased during the recent financial crises. Moreover, the probability of loans to be secured by real or personal guarantees increased with firm riskiness.

Secondly, the paper estimates a multilevel model in which the dependent variable, i.e. the spread between bank loan and overnight interest rates, is regressed on loan-contract, individual-firm and -bank characteristics, and on time dummies.

Results show that collateral helps reduce loan interest rates charged to small-sized businesses before and during the financial crisis. As for personal guarantees, the findings show a positive effect on (i.e. a decrease in) interest rates only during the financial crisis. Indeed, in normal times, personal guarantees do not give the lender a specific claim on particular assets, and restrict the actions (s)he could take in case of borrowers' bankruptcy. Therefore, they have no effect on loan interest rates. During troublesome periods, such as the current financial crisis, providing personal guarantees together with collateral is a sign of borrowers' quality that has positive effects on interest rates.

References

Barba Navaretti G, Bugamelli M, Ottaviano G, Schivardi F (2010) The global operations of European firms. Efige Policy Brief

Calcagnini G, Favaretto I (2011) The economics of small businesses. An international perspective. Springer, Heidelberg

de Nardis S, Pappalardo C (2010) Imprese italiane nella competizione internazionale: modalità di un aggiustamento a lungo misconosciuto, paper presented at the 51st Italian Economists Association (SIE), Catania

ISTAT (2010) I Conti nazionali, http://www.istat.it/it/archivio/43009

Mayer T, Ottaviano GIP (2007) The happy few: the internationalization of European firms. New facts based on firm-level evidence. Brueghel Blueprint 3, Brussels

Rey GM, Varaldo R (2011) Crescita economica, internazionalizzazione e rinnovamento imprenditoriale in Italia. Econ Ital (3):749–789

World Bank (2011) Doing business 2012: doing business in a more transparent world. http://www.doingbusiness.org/reports/global-reports/doing-business-2012

Contents

Contributors

Sergio Arzeni OECD Centre for SMEs, Entrepreneurship and Local Economic Development (CFE), Paris, France, sergio.arzeni@oecd.org

Bruno Basalisco Imperial College London, London, UK, brunix@yahoo.com

Giorgio Calcagnini Dipartimento di Economia, Società, Politica, Università di Urbino “Carlo Bo”, Urbino, Italy, giorgio.calcagnini@uniurb.it

Marco Cucculelli Dipartimento di Economia e Scienze Sociali, Università Politecnica delle Marche, Ancona, Italy, m.cucculelli@univpm.it

Lucia Cusmano OECD Centre for SMEs, Entrepreneurship and Local Economic Development (CFE), Paris, France, lucia.cusmano@oecd.org

Fabio Farabullini Banca d’Italia, Rome, Italy, fabio.farabullini@bancaditalia.it

Ilario Favaretto Dipartimento di Economia, Società, Politica, Università di Urbino “Carlo Bo”, Urbino, Italy, ilario.favaretto@uniurb.it

Annalisa Ferrando European Central Bank, Frankfurt, Germany, annalisa.ferrando@ecb.int

Germana Giombini Dipartimento di Economia, Società, Politica, Università di Urbino “Carlo Bo”, Urbino, Italy, germana.giombini@uniurb.it

Giuseppe Gramigna U.S. Small Business Administration, Washington, DC, USA, giuseppe.gramigna@sba.gov

Donato Iacobucci Dipartimento di Ingegneria Informatica Gestionale e dell’Automazione, Università Politecnica delle Marche, Monte Dago, Ancona, Italy, d.iacobucci@univpm.it

Traci L. Mach Board of Governors of the Federal Reserve System, Washington, DC, USA, traci.l.mach@frb.gov

Jonathan Potter OECD Centre for SMEs, Entrepreneurship and Local Economic Development (CFE), Paris, France, jonathan.potter@oecd.org

E.J. Reedy The Ewing Marian Kauffman Foundation, Kansas, MO, USA, ereedy@consultant.kauffman.org

Giudo M. Rey Istituto di Management, Pisa, Italy, g.rey@sssup.it

Rosalba Rombaldoni Dipartimento di Economia, Società, Politica, Università di Urbino "Carlo Bo", Urbino, Italy, rosalba.rombaldoni@uniurb.it

Robert J. Strom The Ewing Marian Kauffman Foundation, Kansas, MO, USA, rstrom@kauffman.org

John D. Wolken Board of Governors of the Federal Reserve System, Washington, DC, USA, john.d.wolken@gmail.com

Salvatore Zecchini Università di Roma "Tor Vergata", Rome, Italy, drsrzecchini@hotmail.it

SME and Entrepreneurship Policies After the Crisis

Sergio Arzeni, Lucia Cusmano, and Jonathan Potter

1 Introduction

The 2008–2009 financial and economic crisis was the most severe in decades and its costs have been enormous. GDP contracted by about 3.5% in the OECD area as a whole in 2009 and unemployment reached a post-war high of close to 9% on average (OECD 2010a). The longer-term legacy of the crisis is also heavy. Public debt in OECD countries increased to about 100% of GDP at the end of 2011, some 30% points higher than before the crisis. OECD countries may have lost about 3% of potential output. The recovery that seemed to be underway in many OECD countries in 2010 has proved uneven. In several OECD economies, earlier improvements in the labour market have been fading in the course of 2011, and there appear to be greater risks that high unemployment could become entrenched. Economic growth slowed down also in non-OECD countries, including China, where manufacturing production weakened in the first half of 2011. In 2012, world GDP is projected to increase by 3.4%, whereas across OECD countries GDP is projected to rise by 1.6% (OECD 2011a, 2011b).

The crisis had a strongly negative impact on real economic performance of young firms and small and medium-size firms (SMEs). SMEs have been disproportionately affected compared to large firms by the credit tightening and the drastic drop in demand for goods and services, forcing many into bankruptcy and contributing to record levels of unemployment. Indeed, in some countries, SMEs have been shedding jobs at a faster rate than in past recessions, and, on a sectoral level, some categories of SMEs have been hit harder than others, as in the case of export-oriented manufacturing segments, or new firms looking for new credit.

S. Arzeni (✉) • L. Cusmano • J. Potter
OECD Centre for SMEs, Entrepreneurship and Local Economic Development (CFE), Paris, France
e-mail: sergio.arzeni@oecd.org; lucia.cusmano@oecd.org; jonathan.potter@oecd.org

G. Calcagnini and I. Favaretto (eds.), *Small Businesses in the Aftermath of the Crisis*, Contributions to Economics, DOI 10.1007/978-3-7908-2852-8_1,

As it becomes evident that strengthening SMEs and reviving entrepreneurial dynamics are crucial for a rich-job recovery, the market and institutional failures that affect SMEs and entrepreneurs have gained prominence in the policy debate. Governments have been responding to the urgent need of avoiding hysteresis and preserving the fabrics of small viable businesses by targeting their cash flow problems and enhancing access to credit, through extensive use of direct loans or guarantee schemes. In many instances these measures were needed to support otherwise viable businesses and preserve jobs. However, as the fiscal consolidation reduces the room for maneuver, new policy approaches are needed to address long-standing challenges and pursuing the long-term objective of sustainable growth.

Moving from the impact of the recent crisis on SMEs and entrepreneurs across OECD and non-OECD countries, the present chapter discusses the policy environment that has emerged through the crisis and the policy challenges ahead. It highlights the opportunities for economic restructuring and argues for a central role of SMEs and entrepreneurship policies in the future economic agenda. In doing so, it points at key policy areas that have been emerging from analysis and policy dialogue through the *OECD Bologna Process on Entrepreneurship and SME policies* and concludes with the identification of fundamental steps to unleash entrepreneurship.

2 Impact on SMEs of the Financial and Economic Crisis

The recent global financial and economic crisis has had a disproportionate impact on SMEs and entrepreneurial activities and performance across OECD and non-OECD economies. In virtually all countries, the majority of SMEs have experienced a drastic drop in demand for goods and services and a tightening in credit terms, which have severely affected their working capital. The OECD Scoreboard on SME and entrepreneurship finance, which monitors the country level trends and conditions in access to finance, through a number of debt, equity and financing framework condition indicators, shows that, in a large number of countries, business loans and SME loans declined markedly during the recession and, while they recovered somewhat in 2010, they did not reach their 2007 levels (OECD 2012).

Faced with declining sales prospects and credit constraints, most SMEs cut spending by running down inventories, postponing investment and/or reducing staff (OECD 2010b). Nevertheless, bankruptcies increased in many countries. For instance, between 2007 and 2010, in Italy bankruptcies rose from 11.2 per 10,000 enterprises to 20.3. Over the same period, insolvencies nearly doubled in the Unites States and The Netherlands, and increased by two and a half times in Denmark (OECD 2012). In several countries the SME cash flow difficulties were increased by late payments, particularly by public administration and large firms, including big retailers.

The toll taken on SMEs and the slowing down in entrepreneurial dynamics contribute to explain why the crisis brought about the most rapid and sizeable

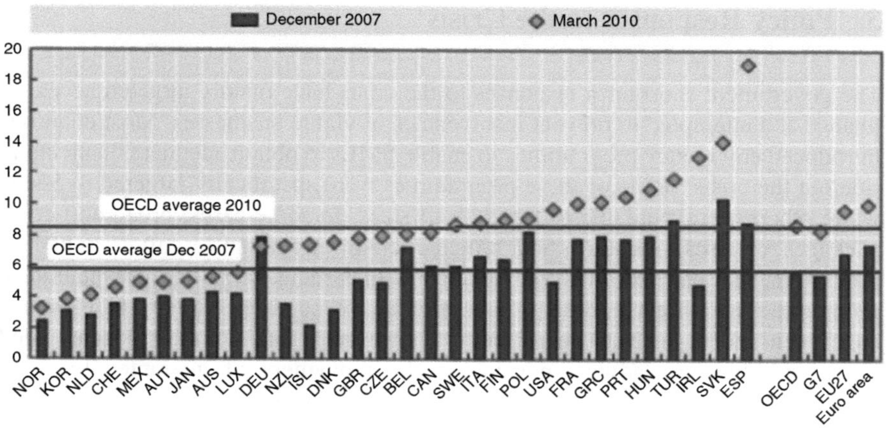

Fig. 1 Unemployment across the OECD, 2007–2010 (Source: OECD Main Economic Indicators 2010)

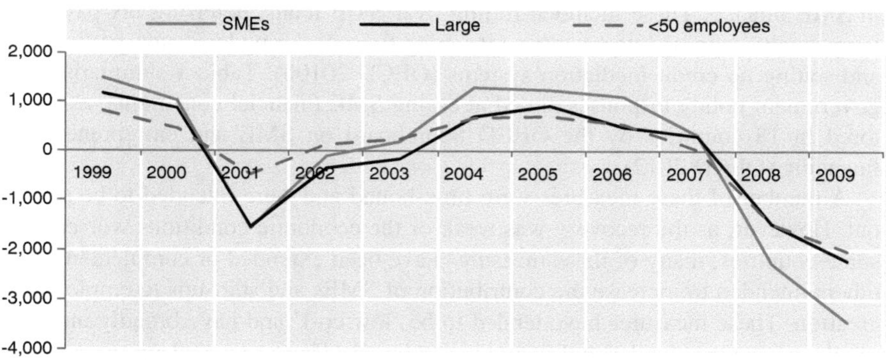

Fig. 2 Net job creation in the United States, by firm size, 1999–2009 (seasonally adjusted, in thousands) (Source: US Bureau of Labor Statistics). Note: "SME" refers to firms with 1–499 employees, "large" refers to firms with 500+ employees

increase in unemployment in the post-war period. The OECD-wide unemployment rate rose to a high of 8.8% in December 2009 from a 25-year low of 5.6% in late 2007. Unemployment increased by 50% on average between 2007 and 2010 (Fig. 1). However, while labour market conditions deteriorated almost everywhere, the extent of the contraction and the form it has taken differ considerably. In the United States, for instance, the recession has triggered unusually large job losses, despite some reduction in average hours worked (Guichard and Rusticelli 2010). In particular, US SMEs shed jobs faster than large firms (Fig. 2), whereas, as it had happened in past recessions, in several EU countries SMEs retained employees more than large firms.

3 Policy Responses to the Crisis

The government short term responses to the crisis have mainly targeted access to finance and attempted to prevent insolvencies of viable businesses. Most countries introduced emergency programmes to enable SMEs to obtain adequate financing to ride out the crisis. Some of these programmes were specifically targeted to SMEs while others were available to firms of all sizes. In some cases, broad anti-crisis packages of fiscal stimulus and targeted credit programmes, often linked to programmes for bank bailouts, were launched. In particular, existing loan guarantee programmes and direct lending programmes have been ramped up or new elements have been added, such as guaranteeing short-term loans, combining guaranteed loans with business advice, postponing the repayment of guarantee loans or increasing co-financing by public agencies and banks, among others. Some governments also intervened in the private equity market, which shrinked in most countries at the height of the crisis, to favour long-term investment and balance sheet restructuring (OECD 2010b, 2012).

Some emergency responses were especially tailored to remedy the deterioration in SME finance. These included rolling over SME loans, deferring tax payments temporarily, capping interest rates, refraining from declaring loans non-performing, and setting up credit mediation systems (OECD 2010b). Table 1 summarises the government policy responses aimed at easing SME financial constraints, as monitored in 18 countries by the OECD Scoreboard on SME and entrepreneurship financing (OECD 2012).

A number of these measures were time-bound and were intended to be phased out. However, as the recovery was weak or the economic conditions worsened in some countries, many of these measures have been extended or complemented by others intended to increase the contribution of SMEs and start-ups to employment creation. These measures have tended to be ‘low cost’ and have broadly included regulatory reforms and simplification, continued efforts to speed up government administration payments to firms, provision of business advisory services, tax breaks, and small equity capital funds to support entrepreneurship. A small group of OECD countries, including Australia, Germany and New Zealand, have either completely removed their crisis measures or did not implement extensive crisis support for SME finance to begin with.

These short term measures have helped SMEs to overcome difficult conditions. However, for growth to be restored and based on more sustainable terms, structural improvements are needed. In some cases it may still be important to assist incumbent firms to survive in order to preserve existing jobs, but, looking to the future, the challenges are likely to be of a different order, also in the light of the fiscal consolidation ahead.

In fact, the stimulus measures have left public debt and deficits at record highs, and in some regions, particularly in Europe, financial markets reacted strongly to this by rising interest rate spreads on sovereign debt. Furthermore, the measures taken to facilitate SME financing, though necessary, have significantly shifted risk from the private to the public sector and increased reliance on debt.

Table 1 Government policy responses to improve SME access to finance during the 2008–2009 crisis

Policy response	Countries
Increased amount government loan guarantees and/or % guaranteed, number of firms eligible, countercyclical loans	Canada, Chile, Denmark, Finland, France, Hungary, Italy, Korea, The Netherlands, Portugal, Slovakia, Slovenia Switzerland, Thailand, United Kingdom, United States
Special guarantees and loans for start ups	Denmark, The Netherlands
Increased government export guarantees	Denmark, Finland, The Netherlands, New Zealand, Portugal, Sweden, Switzerland
Government co-financing	Sweden
Increased direct lending to SMEs	Chile, Hungary, Korea, Slovenia
Subsidised interest rates	Portugal, Thailand
Venture capital, equity funding and guarantees	Canada, Chile, Denmark, Finland, France, Italy, The Netherlands, New Zealand, Sweden, United Kingdom
New programmes: business advice	Denmark, New Zealand, Sweden
Tax exemptions, deferments	France, Italy, New Zealand, Sweden
Credit mediation	France

Source: *Financing SMEs and Entrepreneurship: An OECD Scoreboard* (2012)
Note: The countries monitored are: Canada, Chile, Denmark, Finland, France, Hungary, Italy, Korea, The Netherlands, New Zealand, Portugal, The Slovak Republic, Slovenia, Sweden, Switzerland, Thailand, The United Kingdom and The United States

Over the near future, high levels of government debt have to be reigned in, necessitating in some cases abrupt austerity measures, which strongly limit the room for maneuvering traditional fiscal policy. Against such background, guiding a transition from a policy-driven recovery to self-sustained growth poses several challenges to policy makers and governments, which have to provide a credible medium-term framework, including for the financial sector, to stabilize expectations, strengthen confidence and boost investments, as well as respond to emerging new societal needs.

4 SME and Entrepreneurship Policies in a Post-Crisis Environment: The Challenges Ahead

Fostering entrepreneurship and small business development is a key ingredient to any development strategy, be it in developed or in developing countries. The recent crisis and its effects on the SME sector and potential entrepreneurs have, if possible, highlighted further the crucial role of small businesses and start-ups in the creation or preservation of employment, as well as in the generation and diffusion of innovation. Many empirical studies have shown the aggregate relationships between entrepreneurship and SME activity and economic growth and job creation. These growth and job creation effects happen through innovation, as new firm

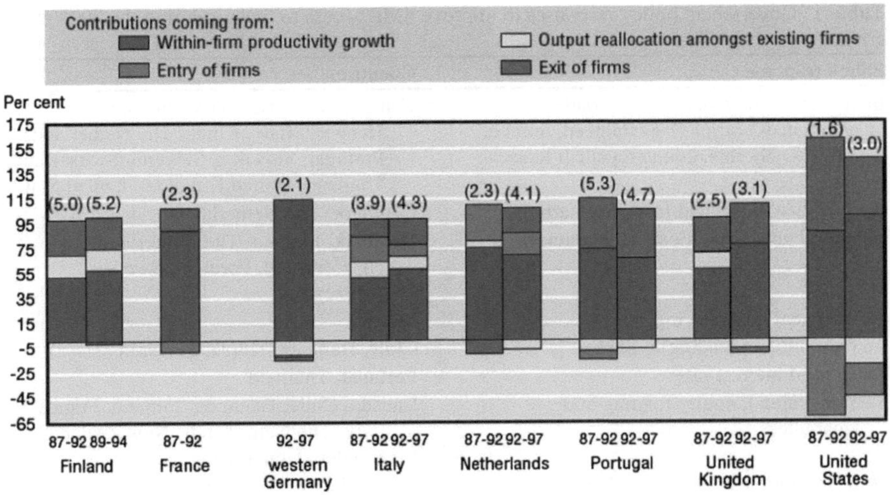

Fig. 3 Decomposition of labour productivity growth in manufacturing, 1987–1997. Percentage share of total annual productivity growth of each component (Components may not add up to 100 because of rounding) (Source: OECD (2003), *The Sources of Economic Growth in OECD Countries*, OECD Publishing). Note: Figures in brackets are overall productivity growth rates (annual percentage change)

creation and SME growth increase productivity, especially in knowledge based sectors, and bring new or under-utilised resources into use (OECD 2010c).

In the new entrepreneurial economy, the contribution of entrepreneurship to growth is much more important than before. Figure 3 presents the decomposition of labour productivity growth in manufacturing sectors for two 5-year intervals: 1987–1992, a phase of expansion for most countries, and 1992–1997, a phase of contraction for most countries. It tells us that within-firm productivity growth accounts for most of labour productivity growth, while output reallocation between existing firms plays a marginal role, if not negative. However, what is interesting from the entrepreneurship point of view is that the combination of the entry and exit of firms (i.e. net entry, which is a proxy for entrepreneurship) is positive in most countries and accounts for between 20% and 40% of labour productivity growth. These estimates suggest that the entrepreneurial process is one of the drivers of growth in modern economies. This is particularly true in periods of slowdown (the 1992–1997 period in the chart), when the contribution of the "entrepreneurship" component (i.e. the net entry element) increases in relative importance, particularly because of the exit of low productive units.

Further, SMEs and start-ups play an important role in limiting social exclusion and engaging disadvantage groups in the economic process. Entrepreneurial opportunities are a formidable channel for economic and social participation and upgrading, and for tapping into talent and innovativeness that flourish in different parts of societies. Entrepreneurship can play an important role by allowing marginalized people, including young people and women, to create their own opportunities to participate in the economy.

Therefore, a key challenge ahead is to set back in motion, on more sustainable basis, the mechanism for enterprise creation, as a driver of job creation and economic renewal. At the same time, over the long-term, policymakers need to address long-standing market and institutional failures with structural reforms, to support the emergence and expansion of new and dynamic enterprises that will carry forward innovation and job creation.

4.1 New and High Growth Firms for Job Creation

New firm creation through entrepreneurship is a key driver of job and wealth generation, especially when the entry dynamics stems from innovation-related opportunities, rather than from high unemployment and uncertain career prospects or, as largely in underdeveloped economies, from necessity (Buenstorf 2009). New and young firms can translate knowledge and ideas into jobs and wealth, often exploiting opportunities that have been neglected by more established companies. However, for entrepreneurial dynamics to sustain innovation, employment and economic expansion, the survival and growth of new firms is crucial. Empirical evidence shows that a great proportion of newly founded firms exit from the market in a rather short time (Geroski 1995) and that survival chances are largely differentiated in the enterprise population, depending on the founder's knowledge endowment, skills and prior experience, among other factors (e.g. Glaeser et al. 1992; Klepper 2001). In other terms, what matters to growth is not only the entry of new entrepreneurs into markets, but also the quality, type, potential and sustainability of the entrepreneurial projects.

The US case is exemplary of the relevance of new enterprise creation for a sustained recovery. In fact, according to the Kauffman Foundation, between 1977 and 2005, nearly all the net job creation (jobs created minus jobs destroyed) in the US occurred in firms less than 5 years old (Haltiwanger et al. 2009). Of the overall 12 million new jobs added to the US economy in 2007, young firms were responsible for the creation of nearly eight million of those jobs (Stangler and Litan 2009). The crisis severely affected this mechanism of job creation and economic renewal.

However, entrepreneurial dynamics and its transmission to market creation and employment generation broadly differ across economic contexts. In the US, for instance, entrepreneurial dynamism is significantly more pronounced than in the European countries, and successful entrant firms appear to enjoy a stronger expansion in their initial years. Nevertheless, also in Europe the contribution of new entrants to employment is non-marginal, although the relevance varies significantly across countries. Figure 4 illustrates, for some European countries, the employment share of employer enterprise births in manufacturing and services. This share is generally higher in the service sector and ranges from around 1% in Belgium to 6% in Hungary, Portugal or the Slovak Republic.

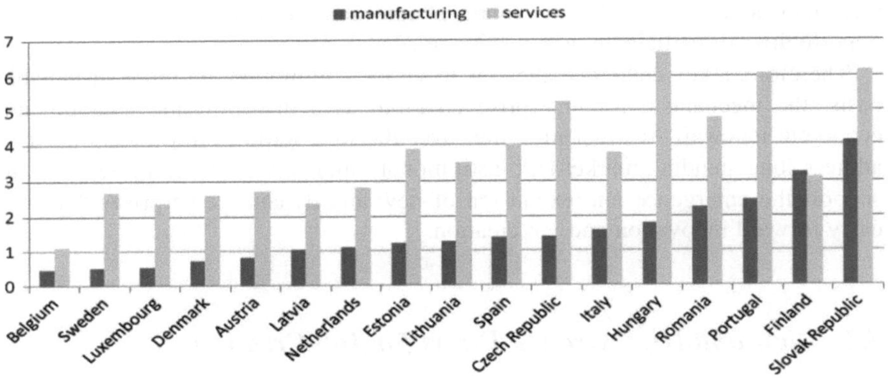

Fig. 4 Employment share of employer enterprises births in manufacturing and services, selected European countries, 2007 (Source: OECD-Eurostat *Entrepreneurship Indicators Programme* 2011). Note: The share is calculated as the number of persons employed in newly born employer enterprises (firms with at least one employee) divided by the number of persons employed in the stock of employer enterprises

Evidence shows that "gazelles" and high growth enterprises[1] play a disproportionate role in job creation, taking into account their relatively small number across all economies (Birch 1987; Birch and Medoff 1994; Birch and Haggerty 1997). According to the OECD-Eurostat Entrepreneurship Indicators Programme, high growth firms represent on average around 3–6% and 8–12% of the total business population, when growth is measured respectively by employment and turnover (OECD 2010d). "Gazelles" are also a very tiny share of the business population, representing on average less than 1% or 2% of firms, when measured in terms of employment or turnover (OECD 2010d). Again, in comparison with the US, where enterprises expand and contract faster, in Europe a larger share of enterprises remains stable over time (OECD 2011c).

In order to generate the jobs of the future, policy should help to identify firms with high growth potential. However, if sustaining the potential for growth is desirable, the a-priori identification of a target population of firms is not straightforward. In the literature, gazelles or high growth firms are frequently associated with innovation and organizational change. However, innovation is not always a prerequisite for firm growth, as firms can grow for a number of other reasons, including the mere utilisation of market opportunities (OECD 2010d). Furthermore, not all fast-growing firms are necessarily new or young (Henrekson and Johansson, 2010). Rather, empirical evidence shows that sustained growth represents a

[1] The OECD-Eurostat Manual on Business Demographic Statistics (2007) recommends defining "high growth firms" as *all* enterprises with averaged annualized growth greater than 20% per annum, over a 3 year period, and with ten or more employees at the beginning of the observation period. Growth is measured by the number of employees and by turnover. "Gazelles" are defined as a sub-group of high growth firms, including enterprises up to 5 year old (OECD 2010d).

transitory phase in the life of an enterprise (OECD 2010d; World Economic Forum 2011). Once started, the majority of SMEs will remain small and appropriate SME policies and framework conditions are needed to maintain the positive effect on employment.

4.2 SMEs and Entrepreneurship for Innovation

Innovation is crucial to long-term economic growth, even more so in the aftermath of the financial and economic crisis, as countries seek to improve productivity performance and to address pressing societal issues, such as climate change, health and poverty (OECD 2010e). In this quest for growth, competitiveness and life quality, however, major changes in innovation policies are needed. The innovation environment has changed significantly over the last decades. Audretsch and Thurik (2004) and Thurik (2009) describe a shift from a "managed economy", centred around the key role of science and systematic large firms R&D, to an "entrepreneurial economy", in which entrepreneurship is one of the foundations of innovation. By serving as a conduit for knowledge spillovers, entrepreneurship links investments in new knowledge and economic growth (Audretsch et al. 2006). New spin-off ventures enable the commercialisation of knowledge that would otherwise remain un-commercialised in large firms, universities and research organisations (OECD 2010c). New and small firms are often active in breakthrough innovations. In some of the most innovative economies, a non marginal proportion of patent applications is filed by young firms, although, as illustrated in much theoretical and empirical literature, differences across technological regimes and sectors are large (e.g. Malerba and Orsenigo 1996; Breschi et al. 2000).

The increasing role of small businesses in innovation dynamics is related to profound and multi-dimensional transformations in market economies, which have led to a reduced importance of economies of scale in production, management, finance and R&D. A major driver of this has been the reduction in the product standardisation that was, on the other hand, the force of large firms in the middle of the twentieth century. As incomes have risen and the taste for variety increased, multiple market niches have appeared, which new and small firms are quick to fill. These include traditional product markets, but also, and increasingly so, social goods and services, such as services for the individual and environmental protection (Noya and Clarence 2007).

Furthermore, changing markets, increased competition and new technologies have reduced product life times, demanding rapidity in anticipating or responding to new market needs. New technologies, such as, among others, micro-electronics, have made it possible for small firms in many industries to produce small batches as efficiently as large firms once produced large batches (OECD 2010c). All this has been associated with what has been termed "flexible specialization", that is, the capacity of firms and economies to specialise – often through outsourcing and fine division of labour at the local level – produce for niche markets and adapt to rapidly changing market environments (Piore and Sabel 1984; Scott 1988; Garofoli 2002).

The ever more important role of small players for innovation is also related to the emerging open or network-based modes of innovation. New and small ventures participate increasingly in knowledge networks, not only as knowledge exploiters, but also as knowledge sources and, increasingly, as "bridges of innovation", which interact with other players as knowledge purchasers, providers and partners (OECD 2011c). This is the case, for instance of Knowledge-Intensive Business Services (KIBS), firms involved in activities such as consultancy, market research, design, engineering and technical services, which provide specialised knowledge inputs for other service and manufacturing firms, generate opportunities for interactive learning, favour the creation of local linkages and contribute to the connectivity of national or regional systems with outside knowledge networks (Muller and Zenker 2001; Miles 2005; Corrocher et al. 2009).

However, it is important to note that there exist an uneven distribution of small firm innovation, between a few highly innovative firms with high growth potential, and the great majority of SMEs that innovate very little compared to their large counterparts (OECD 2010c). While some spin-offs and small businesses have been able to capitalise the advantages from the recent transformations, a large share of SMEs encounter difficulties in adapting to the knowledge economy. In this respect, within-sectoral differences can be more relevant than between-sector variances. In certain high-technology sectors (e.g. semiconductors, biotechnology), emerging sectors (e.g. green industries) and creative industries (e.g. film production, publishing, architecture etc.), innovative SMEs and start-ups have been key players and drivers of innovation, largely based on the combination of intangibles, new technologies and design skills (OECD 2011c). However, in OECD countries, even in traditional sectors, such as household appliances, food industry, paper, wood, furniture, metallurgy and plastics, which may not invest highly in formal R&D, but rather adopt informal and incremental modes of technological and non-technological innovation, SMEs represent between 33% and 50% of innovative firms, continuously creating product, process, marketing and organisational innovations (OECD 2010e).

5 Opportunity for Restructuring and Policy Challenges

As economies gradually emerge from the global financial and economic crisis, policy will be operating in an environment in which significant structural adjustment is expected. High levels of firm creation and growth are likely to take place alongside continued high levels of exits in declining and non-competitive enterprises and activities (Potter and Thompson 2011). The policies to tackle growth and unemployment problems will need to consider a different competitive environment and seize the opportunities for restructuring.

Over the last decades, the role of entrepreneurship and SME development for sustained and inclusive growth have been high in the agenda of the OECD Working Party on SMEs and Entrepreneurship (WPSMEE), a unique international forum to compare policies and identify good practices in fostering entrepreneurship, harnessing SMEs and entrepreneurship for job creation, innovation and growth, helping SMEs meet the challenge of globalization. A milestone in the WPSMEE trajectory has been the launch, in 2000, of the so-called *OECD Bologna Process on Entrepreneurship and SME policies*, by which the OECD brings together over 70 countries and more than 50 international organisations, institutions and NGOs. The objectives are threefold: to foster an entrepreneurial agenda and SME competitiveness at the global level; to provide guidance to governments that will help entrepreneurs and SMEs worldwide reap the benefits of globalisation; and to further deepen and improve the high-level dialogue on SME policies among policymakers, business, national and international organisations and other stakeholders.

The importance of innovative SMEs and entrepreneurship for sustainable development paths has been recognised from the beginning of this policy dialogue. In June 2000, the *Bologna Charter on SME Policies*, adopted by 48 OECD member and non-member economies, recommended improvements to SMEs' abilities to manage innovation, reduction of financial barriers to SME innovation and facilitation of SMEs' access to national and global innovation networks. This was further underlined by the OECD Istanbul Ministerial Declaration on *Fostering the Growth of Innovative and Internationally Competitive SMEs* in June 2004, in which Ministers committed to working co-operatively to achieve progress in reducing barriers to SMEs' access to international markets and assessing the effects of globalisation on SMEs, in particular examining issues related to SME access to financing and to support for innovation.

In November 2010, in the midst of the crisis, policy makers convened at the OECD "Bologna+10" High-level Meeting, to design a roadmap to ensure that SMEs and entrepreneurship make their fullest contribution to the job creation needed for the recovery and to secure sustainable growth over the longer term. The following paragraphs illustrate the main axes of this roadmap.

5.1 Improve Institutional Framework Conditions

The importance of framework conditions in affecting the entry, growth and exit of enterprises has been recognised by Ministers and representatives of governments in the Bologna Charter on SME Policies and the Istanbul Ministerial Declaration. Lack of macroeconomic stability, such as volatile interest rates and exchange rates, reduce investment by SMEs and entrepreneurs. Regulatory environments often impose undue burdens on SMEs and entrepreneurship such as through high and regressive compliance costs, lack of transparency in the application of rules and legislation, inefficient bankruptcy laws and procedures, lack of clarity and coherence in product standards in world markets, unfair or non-transparent competition policy and ineffective anti-corruption measures. The nature of the tax and social security regime has also a potentially powerful influence on entrepreneurial behavior, which may inadvertently discriminate against SMEs and entrepreneurs relative to large existing firms in both effective tax rates and compliance burdens (OECD 2009a).

Innovative SMEs and entrepreneurs also commonly suffer from lack of access to financial services, particularly to seed and development capital, which has been exacerbated by the financial and economic crisis. Indeed, even before the crisis the supply of finance to certain categories of SMEs was affected by significant market and institutional failures (OECD 2006; Potter and Thompson 2011). High growth and innovative firms are characterised by unproved business models (often embodying new technology) that are difficult to evaluate using the traditional credit analysis tools of banking. Almost invariably, the firms will not be able to provide the collateral that banks typically use as a risk-sharing technique. Innovative firms usually have higher rates of failure than traditional firms, making them unsuitable for debt finance. The crisis may have temporarily diverted attention from this long-standing problem, but it is certain to reappear in the post-crisis period, where it will be imperative to identify and finance promising companies.

5.2 Foster Internationalization

New firm start-ups and innovative SMEs increasingly operate in international markets, exposing them to foreign competition and providing them with new opportunities. At the same time, globalisation is increasing the importance of cross-border knowledge flows about markets, suppliers and technologies, which help to upgrade SME competitiveness and stimulate SME growth. Access to knowledge flows is particularly important in the context of the widespread adoption of open innovation methods in many sectors, involving collaboration between new and small firms and other firms and organisations in developing new products and services, new process technologies and new organisational models. Foreign partners can be particularly valuable in such knowledge exchange networks. Indeed, export orientation and active innovation networks are the core characteristics associated with rapid SME growth. However, SMEs are under-represented in the international economy. While they typically contribute around 50% to GDP in national economies, they account for only about 30% of exports, and even less of international direct investment. They are also less likely than large firms to engage in formal collaboration for innovation (OECD 2009b). To some extent, this under-representation is mitigated by the participation of some new firms and SMEs in global value chains, which enables these firms to export and collaborate indirectly as well as directly by supplying other, often larger, firms in their domestic market.

There are many barriers affecting the internationalisation of SMEs and entrepreneurs. Many are internal to the SME, such as limited information on foreign markets and supply chain and technology partners, lack of managerial time for international engagement, lack of workforce skills and knowledge to access markets and collaborators and absorb innovation, shortage of working and investment capital to finance exports and deal with slow supply chain payment schedules and lack of sufficient product and service quality to meet customer requirements. Non-tariff barriers and poorly adapted framework conditions are further problems for SMEs, notably administrative and technical difficulties, exchange rate fluctuations, documentation and payment.

To overcome these barriers, it is important to increase the participation of SMEs in international networks, including international collaborative research programmes, address the financial impediments to internationalization, support SME upgrading programmes and provide brokerage and information services on foreign market and collaboration opportunities.

5.3 *Promote Innovation and Intellectual Asset Management*

The management of intellectual and intangible assets is critical for turning innovation into a driver for SME competitiveness and growth, particularly for new enterprises and SMEs that rely strongly on the exploitation of intellectual capital in their business models. The use of intellectual property rights (IPR) including patents, copyrights and trademarks can be an important tool for protecting and managing intellectual assets, assisting SMEs to open up new markets, increase enterprise value and raise finance, although some firms prefer informal methods to protect their intellectual property such as secrecy and short lead times. However, studies show that SMEs rarely have explicit intellectual assets strategies, lack knowledge of the possibilities offered by IPR regimes and use intellectual property protection to a much smaller extent than large firms (OECD 2011d). A key challenge relates to the internal management practices of SMEs. These are often too driven by short-term thinking based on securing immediate commercial advantage to be concerned with the strategic use of their intellectual assets. This is often coupled with a considerable lack of knowledge about available strategies for formal and informal intellectual property protection.

Further problems relate to the lack of adaptation of current regulatory landscapes to SME operating practices and resources, particularly in the application and enforcement stages. This is even more evident in the case of SMEs operating at the international level, which face high fixed costs, multiple filings, regulatory and technical differences across countries and high uncertainty in enforcement (OECD 2011d).

Policies to strengthen SME innovation and creation of economic and social value out of intellectual assets should, first of all, enhance SME awareness about the strategic opportunities offered by IPRs, facilitate appropriate advice and consultancy on intellectual asset management to SMEs and new firms, but also adapt the IPR system, and its procedures, to the needs and resources of SMEs and entrepreneurs.

5.4 Develop Entrepreneurial Human Capital

Entrepreneurship skills are necessary for organising, leveraging and managing resources for the creation of a new venture. A portfolio of entrepreneurship skills is required including risk assessment, strategic thinking, self-confidence, networking, motivational and other skills. Start-up entrepreneurs often lack skills across a number of the relevant areas. Rather than in the area of small business management skills, such as business planning and accounting, the major gap appears to be in the area of strategic skills associated with entrepreneurship, including decision-making, risk-taking, information processing, opportunity recognition, resource organisation, market awareness and product management. Entrepreneurship education and training programmes in schools, colleges, universities and other organisations seek to address this gap. However, there are significant inequalities across countries in the exposure of individuals to entrepreneurship training (Bosma et al. 2009). Furthermore, there are a number of common weaknesses in entrepreneurship training provision. For example, training is often limited to certain subject disciplines such as business studies rather than reaching all students although there is evidence that graduates from other disciplines often have a greater propensity to start enterprises than business studies graduates. In addition, traditional classroom teaching and assessment methods are often used rather than more practical and interactive methods, while there is commonly a lack of training and support for entrepreneurship teachers.

It is necessary to smartly scale up entrepreneurship education in schools, vocational education and universities, shifting the emphasis from business management skills to strategic skills for growth-oriented entrepreneurship, also introducing interactive teaching methods that incorporate practical experience.

6 Conclusions

Economies are gradually emerging from the most severe global financial and economic crisis in modern times, although recovery is predicted to be a slow and lengthy process for many countries. SMEs and entrepreneurship can greatly contribute to the general recovery, especially in employment terms, and governments should carefully consider them as a key leverage for engineering exit from the crisis and fostering job creation and sustainable growth. Policy now needs to shift emphasis from enabling existing SMEs to weather the crisis to supporting the emergence and expansion of new and dynamic enterprises that will carry forward innovation and job creation in the recovery. The present chapter has discussed key policy areas for long-term growth, as identified at the international level by governments, SMEs and stakeholders through the *OECD Bologna Process on Entrepreneurship and SME policies*. However, as the fiscal environment tightens and sluggish growth carries the risk that high unemployment becomes entrenched, it is urgent that governments adopt immediate measures to unleash entrepreneurship and sustain competitive SME development. To do so, public policies must:

(a) Enforce a drastic reduction in the delay of payments by the public administration and large firms, including big retailers, to small suppliers of goods and services;
(b) Sharply simplify the administrative compliance by small business, as this alone may boost growth by 1%, if effectively implemented;
(c) Support networks of SMEs to get access to public procurement, to expand internationally, particularly in emerging markets, to upgrade the skills of human capital and capture the intellectual assets that are vital for growth.

References

Audretsch D, Thurik R (2004) The model of the entrepreneurial economy. Int J Entrepreneurship Educ 2(2):143–166

Audretsch DB, Keilbach MC, Lehman E (2006) Entrepreneurship and economic growth. Oxford University Press, Oxford

Birch DL (1987) Job creation in America: how our smallest companies put the most people to work. The Free Press, New York

Birch DL, Medoff J (1994), Gazelles. In: Lewis CS, Levenson AR (eds) Labor markets, employment policy and job creation. Westview Press, Boulder, pp. 159–168

Birch D, Haggerty A (1997) Who's creating jobs? Cognetics, Cambridge, MA

Bosma NS, Acs ZJ, Autio E, Coduras A, Levie J (2009) Global entrepreneurship monitor. 2008 Executive report, Global Entrepreneurship Research Consortium (GERA)

Breschi S, Malerba F, Orsenigo L (2000) Technological regimes and Schumpeterian patterns of innovation. Econ J 110(463):388–410

Buenstorf G (2009) Is commercialization good or bad for science? Individual-level evidence from the Max Planck society. Res Policy 38(2):281–292

Corrocher N, Cusmano L, Morrison A (2009) Modes of innovation in knowledge-intensive business services. Evidence from Lombardy. J Evol Econ 19:173–196

Garofoli G (2002) Local development in Europe: theoretical models and international comparisons. European Urban and Regional Studies, 9, pp. 225–239

Geroski PA (1995) What do we know about entry? Int J Ind Organ 13(4):421–440

Glaeser EL, Kallal HD, Scheinkman JA, Shleifer A (1992) Growth in cities. J Polit Econ 100(6):1126–1152

Guichard S, Rusticelli E (2010) Assessing the impact of the financial crisis on structural unemployment in OECD countries. OECD Economics Department Working Papers, No. 767, OECD Publishing. http://dx.doi.org/10.1787/5kmftp8khfjg-en

Haltiwanger J, Jarmin R, Miranda J (2009) Jobs created from business startups in the United States. Kauffman Foundation, Kansas City

Henrekson M, Johansson D (2009), Gazelles as job creators: a survey and interpretation of the evidence. Research Institute of Industrial Economics, Stockholm

Klepper S (2001) Employee start-ups in high-tech industries. Ind Corp Change 10(3):639–674

Malerba F, Orsenigo L (1996) Schumpeterian patterns of innovation are technology specific. Res Policy 25:451–478

Miles I (2005) Knowledge intensive business services: prospects and policies. Foresight 7(6):39–63

Muller E, Zenker A (2001) Business services as actors of knowledge transformation: the role of KIBS in regional and national innovation systems. Res Policy 30:1501–1516
Noya A, Clarence E (eds) (2007) The social economy: building inclusive economies. OECD Publishing, Paris
OECD (2003) The sources of economic growth in OECD countries. OECD Publishing
OECD (2006) The SME financing gap, vol I & II. OECD Publishing
OECD (2009a) Taxation of SMEs: key issues and policy considerations. OECD Publishing
OECD (2009b) Science, technology and industry Scoreboard. OECD Publishing
OECD (2010a) OECD economic outlook 87, May 2010, OECD Publishing
OECD (2010b) Assessment of government support programmes for SMEs' and entrepreneurs' access to finance in the global crisis, OECD, Paris
OECD (2010c) SMEs, entrepreneurship and innovation. OECD Publishing
OECD (2010d) High growth enterprises, what governments can do to make a difference. OECD Publishing
OECD (2010e) The OECD innovation strategy: getting a head start on tomorrow. OECD Publishing
OECD (2011a) OECD economic outlook 90, November 2011, OECD Publishing
OECD (2011b) OECD economic outlook. Interim assessment, September 2011, OECD, Paris
OECD (2011c) Entrepreneurship at a glance 2011. OECD Publishing
OECD (2011d) Intellectual assets and innovation: The SME dimension. OECD Publishing
OECD (2012) Financing SMEs and entrepreneurs: an OECD Scoreboard, OECD Publishing
Piore M, Sabel C (1984) The second industrial divide: possibilities for prosperity. Basic Books, New York
Potter J, Thompson J (2011) Government responses to the impact of the global crisis on finance for young and small enterprises. Strateg Change 20:145–159
Scott AJ (1988) Flexible production systems and regional development: the rise of new industrial spaces in North America and Western Europe. Int J Urban Regional Res 12:171–186
Stangler D, Litan RE (2009) Where will the jobs come from? Kauffman Foundation, Kansas City
Thurik R (2009) Entreprenomics: entrepreneurship, economic growth, and policy. In: Acs Z, Audretsch D, Strom R (eds) Entrepreneurship, growth, and public policy. Cambridge University Press, Cambridge, pp 219–249, Ch. 10
World Economic Forum (2011) Global entrepreneurship and the successful growth strategies of early-stage companies. World Economic Forum, New York

OECD Countries' Policy Approach to the SME Crisis in the Recent Recession

Salvatore Zecchini

1 SMEs' Vulnerability

Small and medium sized enterprises (SMEs) are a driving force for growth in any advanced OECD economy, yet a weak component in times of economic slowdown. In the period between 2002 and 2007, SMEs were the most dynamic sector of the EU economy.[1]

In the years 2008–2009, they instead displayed their greater vulnerability than large firms to the adverse economic climate. According to most business surveys, SMEs have been hit hard by the recession, although their output declined less than that of large firms (−5.5% vs.−6.5%).[2] Even in the current phase of gradual recovery, they are lagging behind large enterprises, in spite of the much publicized government support measures that are targeted specifically to them.

The OECD countries' policy response to the SME crisis is the focus of this article. After identifying the causes of SME weakness and the channels through which the recession affected them, the government policy approach is spelled out and its impact is evaluated, leading to a number of conclusions.

What are the main structural sources of SMEs' greater vulnerability? First is their relatively higher sensitivity to variability in sales, due to the relatively narrow range of their goods and services, as well as to the limited scope of their markets, which are mostly domestic or local. Other sources are their lesser market power by comparison with large firms and their frequent position as suppliers to big firms. Furthermore, their relatively lower propensity to engage in R&D and innovation

[1] EIM, pg. 12, 19–20, 28; European Commission, 35; ISTAT; Unioncamere-Infocamere, Movimprese data.

[2] European Commission, pg. 11.

S. Zecchini (✉)
University of Rome "Tor Vergata", Rome, Italy
e-mail: drsrzecchini@hotmail.it

G. Calcagnini and I. Favaretto (eds.), *Small Businesses in the Aftermath of the Crisis*, Contributions to Economics, DOI 10.1007/978-3-7908-2852-8_2,

projects makes it harder for them to adapt to rapidly changing market preferences. Nevertheless, new technology, such as ICT, is a major factor in the continuing increase of small enterprises.[3]

The other major vulnerability is their relatively weaker capital structure.[4] SMEs rely mainly on internal funding sources, since access to credit markets is more difficult for them, while equity capital is generally in short supply. Undercapitalization, a common weakness of SMEs, also curtails borrowing from banks. Yet in some European countries, dependence on short–term bank credit is the only alternative in order to overcome the relative dearth of self-financing.

Of course, SMEs are a highly diverse group of firms. They are lumped together to distinguish them from the large ones, but the criteria for the distinction differ according to country and economic sector. Among SMEs overall, small firms and micro firms are the most vulnerable. Medium-sized firms are more capable of entering foreign markets and investing in R&D. Innovative small firms can grow very rapidly and defend their market shares thanks to intellectual property rights.

2 The Impact of the Recession on SMEs

Regardless of the differences within the group, all SMEs were hit by the recession through several channels, as is shown by the OECD countries' replies to the OECD Questionnaire at the end of 2009.[5] Although their ability to weather the recession varied according to the severity of the recession in the various economic sectors, on the whole they fared worse than their larger counterparts.

Four main channels can be identified: (a) sales; (b) finance; (c) external trade; (d) payment disruption.

The sudden collapse of the demand had a far reaching impact on output, investment, employment and cash flow. As data on these aggregates is rather scanty for SMEs, the main source of information is given by business surveys.

In 2009, SMEs' sale decline in OECD countries varied between 15% and 44%,[6] with a similar decrease in their value added. Another survey,[7] covering Italian firms with more than 20 employees, found that in just two quarters between 2008 and 2009, sales fell by 20% in manufacturing and 14% in services. The slump in market turnover triggered a tendency by large enterprises to bring back in-house a number of activities that had been outsourced, thereby aggravating the SMEs' sales downfall. At the same time, large firms managed to win better price and payment terms

3 European Commission, pg. 23.

4 See among others the latest study by De Socio.

5 OECDa Table 1, OECDb, Table 1. For a general discussion of these issues, see also Zecchini.

6 OECDb, pg. 8–9.

7 Banca d'Italia b.

from their small suppliers.[8] As a consequence, the cost of adjusting large firms' activity to lower demand levels was borne, more than had been expected, by SMEs.

The OECD conducted two opinion surveys among OECD countries' officials, one in early 2009 and another in 2010, about the impact of the recession on SMEs. Their findings [9] signal that albeit these firms experienced a severe contraction of economic activity, the worst affected were those medium-sized, rather than the small ones, as well as those export-oriented, because of the collapse of international trade. Less affected were, instead, fast growing, innovative firms.

Partly as a reflection of this difference, employment in small and micro firms appears to have declined much less than in large and medium sized firms,[10] with the result of a relatively more negative outcome on productivity growth. This trend doesn't seem to be sustainable in the years 2010–2011, as it is expected that with the economic recovery being more lively among large and medium-sized businesses, labour shedding will become relatively more important among small firms. The employment picture appears the reverse in the US, with employment being particularly hit among SMEs.

The second most important factor in the crisis was the curtailment of external financing to most SMEs.[11] This effect cannot be spotted just by looking at indicators of credit demand and supply, but would rather require a complex counterfactual analysis of what the outcome would have been, had the banks and capital markets not experienced the 2008–2009 global financial crisis. As sales and new orders fell, SMEs reacted by cutting down investment, profit margins, and to some extent their labour force. But they had also to face a lower availability of bank credit in the EU until the end of 2009,[12] a sharp tightening of bank credit terms,[13] a risk aversion by financial investors, higher interest cost as compared to large firms, more requests for collateral and guarantees,[14] and a rise in trade payment delays. Their need for working capital financing, therefore, increased significantly and could not be met adequately.[15]

Surveys of SMEs seeking external financing don't provide the full picture of the financial difficulties for the simple reason that they do not adequately account for discouraged borrowers, that do not apply for credit because they presume that their loan application would be rejected. Those surveys actually show relatively small percentages of firms applying for loans in most OECD countries, as well as

[8] EIM IKEI.

[9] OECD b.

[10] European Commission, pg. 12. The relatively lower intensity of labour shedding might also be due to the SMEs' tendency to retain their work force because of the high cost involved in labour training in case of high labour turnover.

[11] As to Italy, see ISAE pg. 70.

[12] ECB b, pg. 2.

[13] ECB a, particularly charts 1 and 2.

[14] OECD f and g.

[15] OECD b, pg. 18, Table 3.

discouraged from borrowing. In the USA, France and Italy loan demand fell in 2009, in the UK only 13% of firms sought new credit. In a phase of deep recession, this is most likely the result of both cutbacks in firms' investment programmes, and their concern for limiting their financial exposure in order to lower default risk.

In the Eurozone, the ECB surveys [16] conducted in the year 2010 also signal a decline of SMEs demand for loans, while the percentage of firms discouraged from borrowing remained broadly stable at 6%. This percentage, however, seems too low in the face of a continued deterioration of the availability of bank loans until the third quarter of 2009.

In any case, a shift in composition of credit demand was evident in most OECD countries. While demand for long-term loans might have declined because of lower investment expenditure, loan applications in 2009 were chiefly aimed at covering short-term working capital needs in the face of severe liquidity shortages, and to a lesser extent, at consolidating short-term debt.

On the credit supply side, the evidence from national financial statistics is rather mixed, but in business surveys, including at the EU level, there are clear signs that loan expansion was seriously affected by supply-side constraints related to banks' new financial conditions and attitudes.[17] Lending conditions were also tightened in both terms and cost [18] for all firms, especially for small ones. Banks required more collateral, tightened loans covenants, asked for higher interest spreads, charged higher lending fees and took longer time to process loan applications.[19] At the aggregate level, bank credit in most EU countries increased slightly or remained flat in 2009, with the notable exception of countries like Italy, where there was a small decline.[20]

In any event, it is clear that in times of a global financial crisis triggered by excessive risk taking and mispricing of credit risk, banks had to lower the risk profile of their asset portfolio, raise their capital base and be more selective in granting credit. This is likely to have resulted in a increasing degree of credit rationing for SMEs, since they are generally undercapitalized, heavily dependent on bank credit, with little access to alternative funding sources, and low market power.

SMEs' conditions were also worsened by the severe downturn in external trade, with exports falling at the recession trough in the second quarter of 2009 by 29% and 23% respectively in Japan and Italy, and less in France and UK (12.6 and 13.4% respectively).[21] To these developments a number of countries reacted by favouring domestic suppliers through various protectionist measures.[22]

---

[16] ECB a.

[17] Hempell and Sorensen.

[18] ECB b, pages 6. 8, and following. For Italy, see also Panetta and Signoretti, pages 22, 43.

[19] OECDb, Table 3. As regards Italy, see ISAE pg. 69–70.

[20] Banca d'Italia a, pg. 180.

[21] OECD c.

[22] According to the EU Commission, 322 new trade restrictive measures were introduced in the world during the recession. See EU Commission e.

An additional consequence of the recession was the rise in trade payments delays, that was particularly evident in Italy, where the days of delays jumped on average from 55 to 63, with public bodies being the worst in payment.[23] But increases on a smaller scale also occurred in other EU countries, prompting the EU Commission to propose a Directive which limits payment delays.

Overall, the combination of the slump in trade, the severe tightening of bank lending, especially for SMEs with a weak capital structure, the drying up of equity sources and the extension of payments delays amplified the impact of the recession on small firms, compounding the risk of a prolonged financial distress and a delayed economic recovery.

3 The Government Approach

The reaction of SMEs to the crisis was to save liquidity and cut costs, investment and margins, but that was just a defensive move to avoid insolvency. It was not enough to engineer their recovery without the Government having to support them.

All OECD countries responded to the SME crisis broadly at the same time and with a remarkable convergence in the policy approach, but with different intensity, depending on the severity of the crisis, and with significant innovations in some countries. In the EU alone, more than 500 policy measures were adopted in favour of SMEs, with the largest share going to address credit access problems.

In general, OECD countries' strategy dealt with the several fronts of the crisis, from the macroeconomic demand management, to the tax burden, need for working capital support, financing of investment and innovation programs, improvements in the business environment, and rescue of relevant firms. Their multifaceted strategy comprised four main segments:

1. Macroeconomic support to aggregate demand, both consumption, investment and employment;
2. Sectoral intervention on the supply side, particularly in the financing area;
3. Structural reforms to make the economic environment more favourable to enterprises;
4. Enterprise crisis management.

The underlying rationale was twofold: (a) restore confidence within the banking industry in order to reopen credit flows; and (b) ease the liquidity squeeze on SMEs in order to allow them to invest and be ready to take advantage of the recovery, when it got under way. The mix of measures varied from country to country with respect to their beneficiaries, scope, duration, and specific focus on SMEs. At the macroeconomic level, the financial policy stance was made exceptionally

[23] Banca d'Italia c.

accommodating, with interest rates reaching record low levels, insolvent banks rescued and restructured, public budget deficits allowed to expand as a consequence of increased social needs and lower tax receipts. In 2008, the EU launched a European Recovery Plan,[24] that envisaged a fiscal stimulus worth € 200 billion (1.5% GDP), of which €120 billion were expected to come from national budgets and the rest from the EU budget and the European Investment Bank.

Still these measures were of temporary nature and aimed at dragging the economy as a whole out of recession. SMEs did benefit from this accommodating stance, since it helped to prop up demand levels, mitigate banks' restrictive attitude and reduce borrowing costs. But they did not benefit as much as large ones, because interest rate spreads widened up drastically for them, as it had been expected in times of intensifying credit selectivity.[25]

Unemployment benefits were provided generously and proved to be essential to support income and demand levels, but small firms were often outside the remit of these measures. Public procurement programs, as usual, did not actually give preference to SMEs, except in the USA, but could anyway turn out to be to their advantage to the extent that they were suppliers to larger ones. In the EU, preference for SMEs in public procurement is not allowed on the very questionable ground that it would infringe competition rules.

In major OECD economies, apart from the massive rescue of insolvent banks, the main policy emphasis was on macroeconomic measures, following the logic that by helping raise the tide, all boats will be lifted, large and small alike. This is particularly the case of Italy, which devoted most of its increase in public expenditure to funding income payments to the newly unemployed. Even so, the amount of public resources committed in the Italian budget to support demand was lower in GDP terms than in other major OECD economies,[26] in view of Italy's precarious state of its public finances. However, a significant effort was made to improve the quality of public expenditure.[27]

Sectoral measures were adopted by the large majority of countries, but few of them targeted SMEs specifically. This should not overshadow that SMEs were also affected by these measures. For instance, the Government's rescue of major banks in state of insolvency or burdened by sizeable non-performing assets, helped restore stability and trust in the financial system and led to an easing of monetary stringency also for SMEs. Another instance: even though tax incentives for the purchase of new cars (cash for clunkers) were introduced in a number of countries with no specific aim at SMEs, but to bolster the important automotive industry, their effects also benefitted the small suppliers to the large car producers. Likewise, public subsidies to investment in R&D and innovation were granted by several countries

---

[24] European Commission c.

[25] As regards Italy, see Panetta and Signoretti, pg. 22.

[26] EU Commission b, Table 1.

[27] Codogno.

to all enterprises to boost their competitiveness and to counter the cuts or discontinuation of these programs by cash-strapped firms. But small innovative firms were best poised to draw on them.

A number of measures were, however, directed to SMEs in the tax, finance, and payments domains.[28] At the European level, in 2008 the EU Council adopted the Small Business Act, for the purpose of creating an institutional framework to stimulate governments to support SMEs on a permanent basis.[29] At national level, a few countries, notably France and Italy, made public funding of banks conditional on their commitment to expand their lending to SMEs. Tax deferrals, simplification of tax compliance, and accelerated depreciation of investment assets were introduced by some of the main OECD countries, such as France, Italy and Canada, with a view to relieving small firms from some administrative burden and alleviating their liquidity shortage.

Under pressure from small firms, few countries, such as France, UK and the Netherlands, shortened payment delays in public procurement contracts and adopted stricter rules for payment compliance among firms. In Italy, the Government did not accelerate payments for public procurement, but offered SMEs the possibility of receiving a public credit guarantee for their claims against a public body, provided the latter certified its debt.

It was in the financial domain actually that governments concentrated their measures aimed at supporting SMEs, particularly in the EU,[30] even though the bulk of their financial commitments was to bail out insolvent banks. Most countries expanded credit guarantees to SMEs for the purpose of inducing banks to reopen their lending facilities, without raising the risk profile of their asset portfolios. The amount of funding was increased substantially, eligibility constraints were eased, a higher percentage of each loan was guaranteed, and applications were processed more rapidly. Guarantees applied both to working capital loans and investment projects.

In parallel, as foreign trade shrank, export credit support was enhanced. Export credit facilities were boosted, their terms were eased, and additional resources were committed for export credit insurance. Some countries, like Canada, also enhanced their direct official lending programmes. In a number of countries, some forms of hidden protectionism also crept up with a view to giving a preference to domestic producers over foreign competition. For instance, some governments pressed multinational corporations to carry out any downscaling of production at foreign plants first, before touching domestic output and jobs. Incentives were also granted to bring outsourced production back home.

As public credit guarantees, however significant, were not enough to prompt banks to expand their lending to SMEs to the expected extent, some countries

[28] OECD a, Table 3.

[29] European Commission d.

[30] European Commission, pg. 5; OECD, Table 4.

resorted to other tools. For instance, rescue packages for banks under stress included a form of conditionality, whereby the rescued banks had to increase credit flows to SMEs. In France and Ireland, this commitment was more constraining than in other countries. In Italy, the constraint was not quantified in a specific rate of increase, but involved only a generic pledge to lend more to SMEs.

Another instrument designed for the same purpose was *credit mediation,* which was introduced by France and Belgium and adopted by Germany since 2010, to overcome banks' reluctance to lend to SMEs. Small firms that were denied a credit application could ask for the intervention of an official credit mediator, that would convene the potential lender for a tripartite negotiation, involving the bank, the public official and the entrepreneur. In Italy and USA, instead, a monitoring system on credit flows to SMEs was established. In Italy, SMEs could file a complaint with a public official, that would call on the bank to clarify its position. No outright credit mediation takes place in this context, but there is a sort of public sponsorship of negotiations between the bank and the potential borrower.

Much more effective were Italy's government-sponsored *debt moratorium* and the collective credit agreements between major Italian banks and the most important SME associations. Under the moratorium, with the bank's agreement SMEs could delay the repayment of medium-term bank loans for 1 year without penalties, while under the collective agreements, banks pledged to cooperate with SME associations to devise appropriate solutions to SMEs' credit difficulties. In Finland, a public financial company, Finnvera, revived an old instrument, the countercyclical loan, to fund recession-hit SMEs on a 5–7 year basis, provided that in parallel, their creditor bank was willing to reschedule their debt for at least 1 year. The purpose is to increase lending to SMEs, rather than replacing private with public lending.

Other innovative instruments included Italy's on-lending program and its promotion of investment in SMEs' equity to cope with the undercapitalization problem. Few other countries, such as France, adopted the same approach. Italy's on-lending involved a public fund (€ 8 billion) lending to banks, that had in turn to on-lend to SMEs at a favourable rate.

Government investment in SME equity and the promotion of SME recapitalization took various forms, particularly participation in new investment funds and private equity investments targeted to SMEs. Only few countries undertook such measures, such as Italy [31] and France, which leveraged the resources of the *Cassa Depositi e Prestiti* and *Caisse des Dépôts*, respectively, together with those of major banks. In the UK, instead, tax benefits were granted to investors in private equity and venture capital. Italy also granted venture capital investors a limited tax exemption on their capital gain.

Within the realm of public financing, the European Investment Bank (EIB) committed sizeable resources in favour of SMEs. The EIB group planned to lend

[31] The initial investment was € 1.2 billion, and is supposed to go up to € 3 billion in the coming years.

€ 30 billion to EU-based SMEs in the period 2008–2011, and has already signed loan agreements for € 20.8 billion, of which € 2467 million for Italian firms. These resources are lent to EU member states' banks for on-lending to small firms at a cost close to that of prime borrowers, i.e., much lower than what SMEs would have to bear in the financial market because of their relatively higher risk.

In the area of structural reforms, few initiatives have been undertaken, but with important potential effects. Tax simplification, cutting down red tape for the establishment of enterprises and simplification of EU state aid rules[32] were the most prominent interventions. There have also been other significant measures, if more limited. Italy granted tax benefits to SMEs that sought stock exchange listing, or engaged in alliances with other firms to develop value chains or synergies. Moreover, it simplified business start-up requirements to promote entrepreneurship. Some countries eased bankruptcy procedures in order to give entrepreneurs a second chance.

An important part of the Governments' policy response was to deal on a case by case basis with the crisis that hit hundreds of large and medium-sized firms, especially in less developed regions. Italy and France applied ad hoc interventions under different forms, ranging from capital injections into large corporations to funding of enterprise restructuring plans and unemployment benefits, and assistance in finding new investors.

Overall, in most countries the government approach to SMEs was focused on two priorities: first, macroeconomic support to newly unemployed workers and demand levels, and second, restoration of normal conditions in the financial system together with easing SMEs' credit access to bank credit. This strategy was pursued through standard tools, such as public budget "automatic stabilizers" and credit guarantees, with few innovative measures and structural reforms. The urgency to stop economic activity from sliding further down as a result of banks reducing their financial exposure prevailed over the need to improve the business and financial environment through radical reforms. SMEs did obtain preferential treatment in some areas particularly in the tax, financial and export domains, but what mattered most to them was public financial support to reopen credit access. After all, these firms did not receive significant privileges over the others, but only assistance once it had become evident that they were bearing the brunt of larger firms' adjustment cost.

Italy's approach differs somewhat from that of other countries to the extent that more room has been devoted to innovative and structural measures in favour of SMEs. One reason was the need for more extensive intervention in a system that was already under stress before the recession. Another was the little room it had on the public budget front, with the consequence that structural measures implying budgetary costs had very limited scope. Structural innovations included government financing of equity and quasi-equity instruments, benefits for enterprise

[32] European Commission d.

recapitalization, a private equity fund for SMEs, subsidies for technology transfer from research to business sector and for start-ups, administrative simplification for starting a business, legal recognition of inter-firm network agreement, urban free zones, funding for on-lending to SMEs. Only France has matched Italy in innovative actions, and has been able to enact the new tools more rapidly and to commit larger amounts of resources.

4 The Impact of Government Measures

Apart from releasing the uptake of some measures, countries have shown little interest in assessing the impact of their policies on the four channels through which the recession hit SMEs. Only few countries have carried out a review of their assistance programs, while the OECD canvassed member countries and beneficiary enterprises about their assessment through two questionnaires.[33] The European Commission also conducted a first review of the implementation of the Small Business Act (SBA) by member countries, gaining information on the support provided to SMEs under a different policy heading. According to the EU Commission, the SBA allowed more than 110,000 SMEs to obtain funding and 200,000 firms to get bank loan guarantees. It also led to the reduction of average cost and time for setting a business start-up from 12 days and €485 in 2007 to 7 days and €339 in 2010.[34]

Albeit any evaluation at this stage is still partial, some evidence is already available, though it is more based on macroeconomic figures and the results of opinion polls than detailed statistics on actual policy results. Econometric estimates of policies' economic impact are also lacking, except for an handful of studies using a counterfactual methodology, such as Caivano et al. on Italy,[35] Blinder and Zandi [36] on the US economy and Canada's Department of Finance.[37] Most of the evidence on SMEs, instead, pertains to the uptake of support programmes, financial conditions, views on Government support and foreign trade results.

On macroeconomic grounds, it is clear that demand management policies succeeded in engineering an upturn in economic activity by mid-2009 in most countries, with Germany and France economies rebounding somewhat earlier (Fig. 1). Output gap rates, i.e. deviations of actual GDP from potential one turned

[33] OECD b, c, d. A first assessment was carried out among government officials in the last two months of 2009, and was updated in July- August 2010, while the inquiry among beneficiary enterprises was made in August-September 2010.

[34] Declaration of the EU Commission Vice President, A. Tajani, on May 25, 2011, reported in the EU Commission website.

[35] OECD d.

[36] Blinder and Zandi.

[37] See http://www.fin.gc.ca/activity/pubs/eval-eap-pae-02-eng.pdf

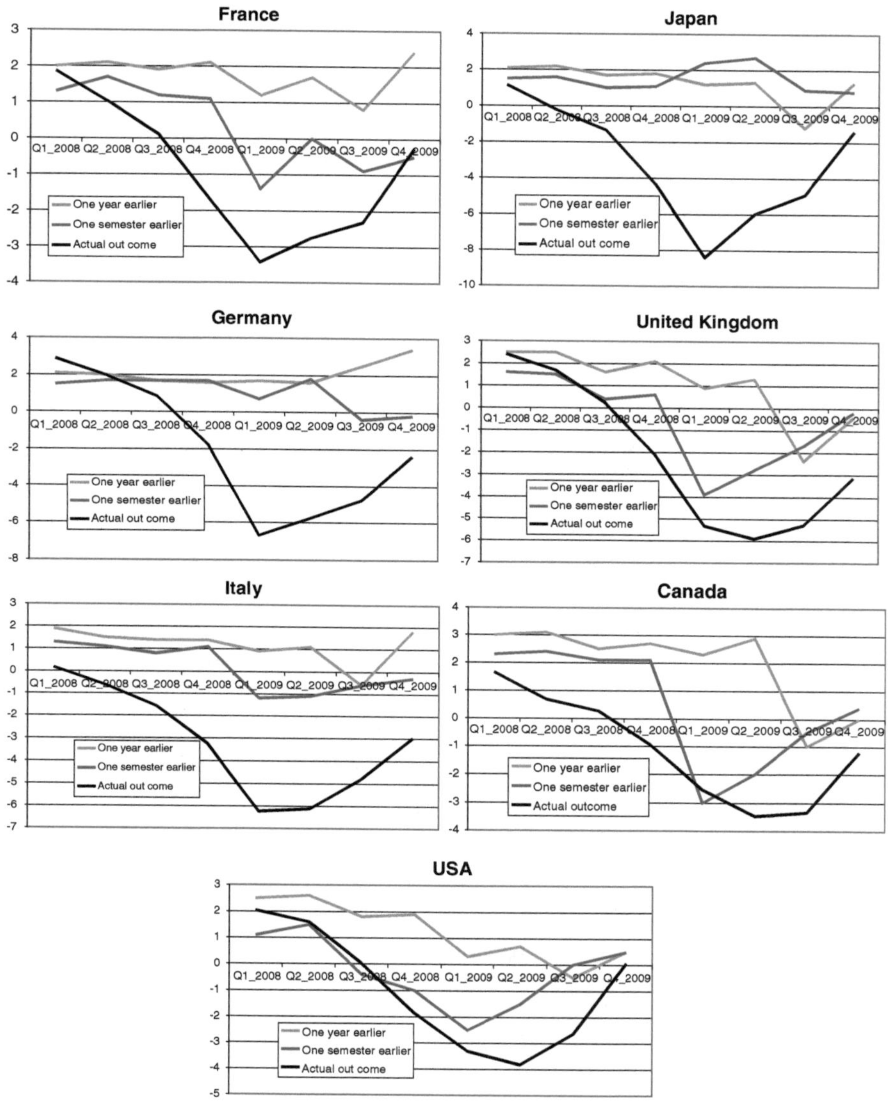

Fig. 1 Real GDP growth and projections (Data Source: OECD Economic Outlook)

deeply negative only in 2009, while in 2008 they were positive or close to zero, with the exception of the USA and Italy (respectively,−0.9 and−0.6% according to OECD estimates) [38] (Fig. 2).

[38] OECD e.

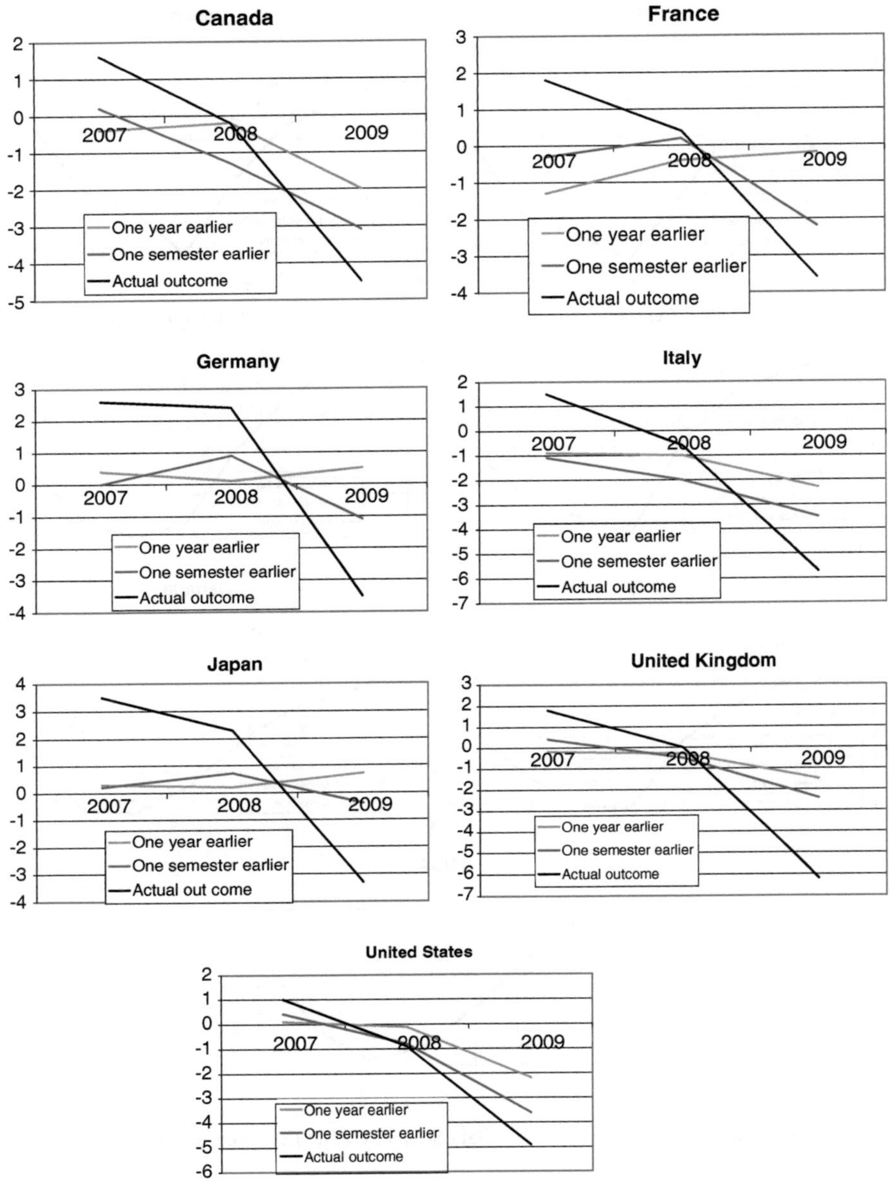

Fig. 2 Output gaps and projections rates (Data Source: OECD Economic Outlook)

A closer look at quarterly GDP growth can shed light on the policy impact, if these data are compared with a counterfactual baseline trend. Caivano et al. carried out such an exercise for Italy, and estimated that in 2008–2010 the cumulative effect of the recession on real GDP growth rate would have been a reduction by about 10% points compared with a baseline trend. But the adopted monetary and financial policy measures offset about 3.5% points of the loss. For the US, Blinder and Zandi estimate that the offsetting impact of Government measures came to 11.5% of GDP and 8.5 million jobs saved, as compared with the recession trend line. For Canada, the Economic Action Plan is believed to have mitigated the deepening of the recession by 4.9% points over the last three quarters of 2009 and to have improved the employment situation by 2.1 points. Of course, the validity of these estimates is predicated on the reliability and stability of the model's parameters, as well as on the realism of the underlying assumptions.

Some inference can also be drawn from a comparison of quarterly developments in macroeconomic aggregates with their ex-ante projections, provided that the latter incorporate the effects of implemented policy measures. This comparison is shown in Figs. 1, 2, 3, 4, and 5 for GDP growth, output gap, consumption, unemployment and exports of goods and services, the most important impact channels for SMEs. For this comparison, a benchmark is drawn from the OECD projections incorporating the expected macroeconomic effects of Government measures. Such a benchmark is derived from simulations based on the OECD econometric model, supplemented by the experts' best judgement.

Accordingly, it can be seen that deviations of GDP quarterly growth from 1-year-earlier projections widened until the first quarter of 2009 for all the G-7 countries, with actual output trend always remaining below the projected one. From then on, deviations have narrowed gradually. For Italy, the deviations were larger than for the other G-7 countries both at the beginning and at the end of the 2008–2009 period; the narrowest deviations were observed for USA.

The same pattern is seen for export growth, but not for real private consumption, which is more varied. In the US, consumption growth deviated downward less than in the other G-7 countries, and actually matched projections in the last quarter 2009. In Japan and Canada, it exceeded projections only in last quarter of the period under consideration, while in Germany and France, it did so since the first quarter 2009. In Italy, real private consumption grew less than projected for all the quarters. This evidence is consistent with the greater concern of the US than the other G7 countries to bolster domestic consumption.

Unemployment rates largely exceeded projections in 2009, but not in 2008, for all the G-7 countries, except for Germany, where it was broadly in line with expectations of a lagged employment response to the economic recession, possibly thanks to a more effective labour market policy.

How to interpret these deviations? They can be read as the product of several factors, including the OECD ability to estimate ongoing trends. But one factor should certainly be the lagged impact of policy measures. OECD projections[39] actually "...illustrate what are the likely outcomes for growth, inflation and employment, given announced policies." Of course, there may be large margins of error around the point projections.

[39] OECD EO Sources – On policy assumptions, from OECD web site.

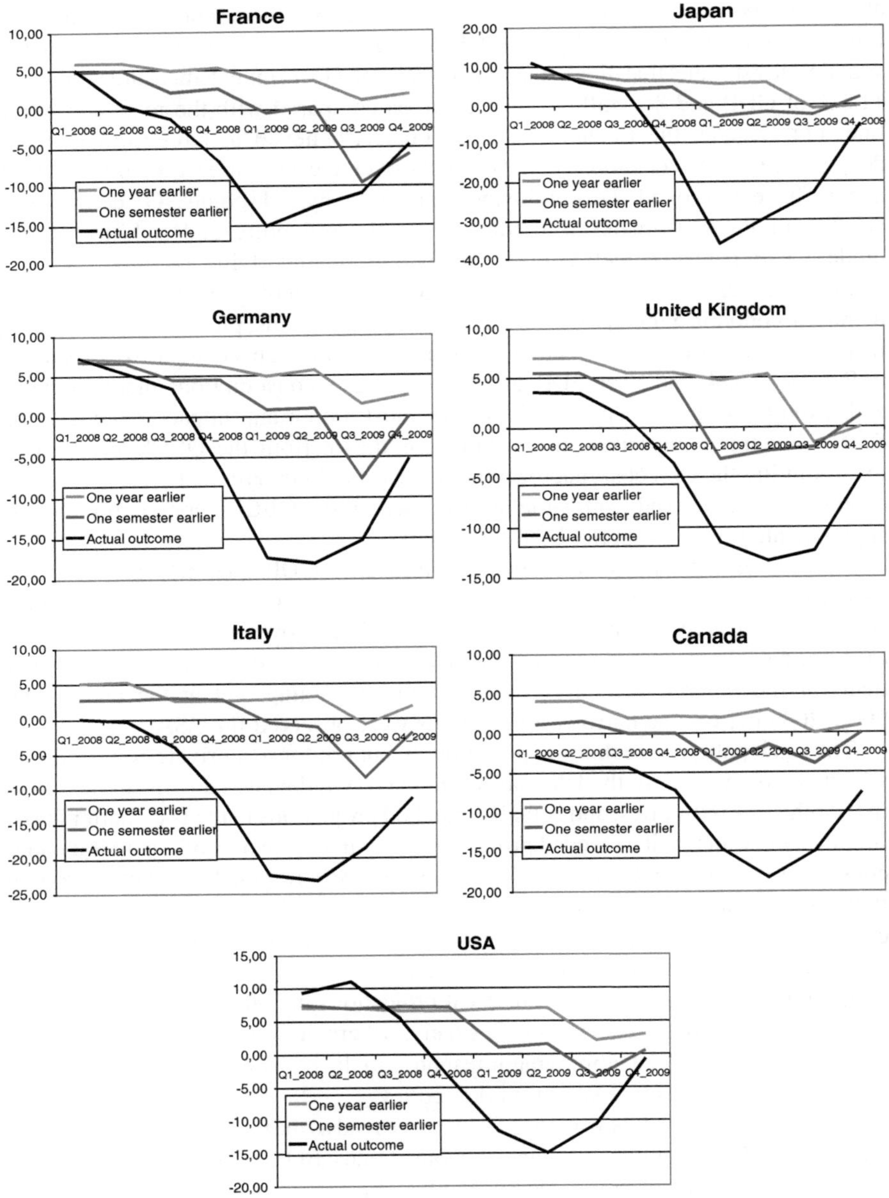

Fig. 3 Exports of goods and services and projections growth rates (Data Source: OECD Economic Outlook)

According to this interpretation, we can surmise that the impact of new measures was by and large insignificant in 2008, but began to be felt in 2009, as evidenced by actual outcomes moving closer to projections for real GDP growth and private

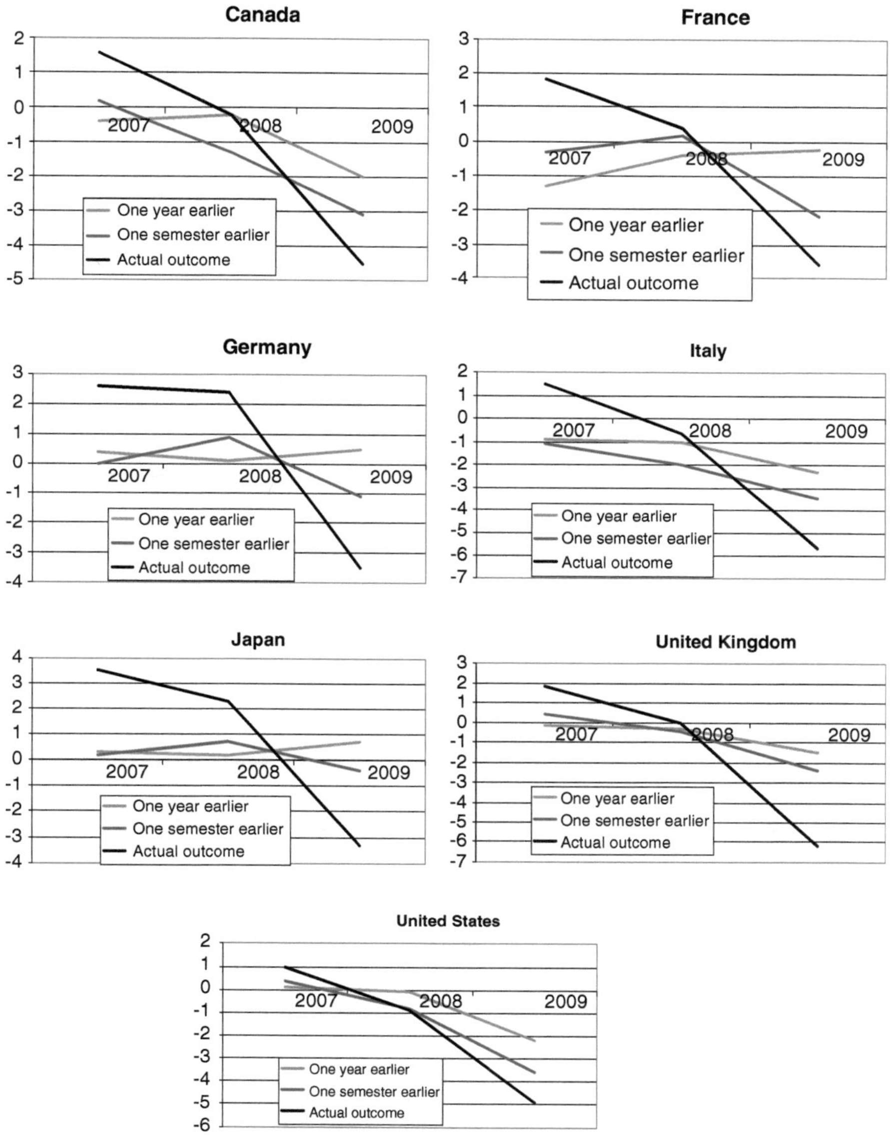

Fig. 4 Private consumption and projections growth rates (Data Source: OECD Economic Outlook)

consumption rates. It might be that in 2008 Governments didn't enact support measures fast enough or strong enough to prevent a deepening of the slump in output. This was reflected in a significant loss of confidence among consumers, with private consumption falling rapidly beyond projections, which were instead discounting a flat trend for 2008.

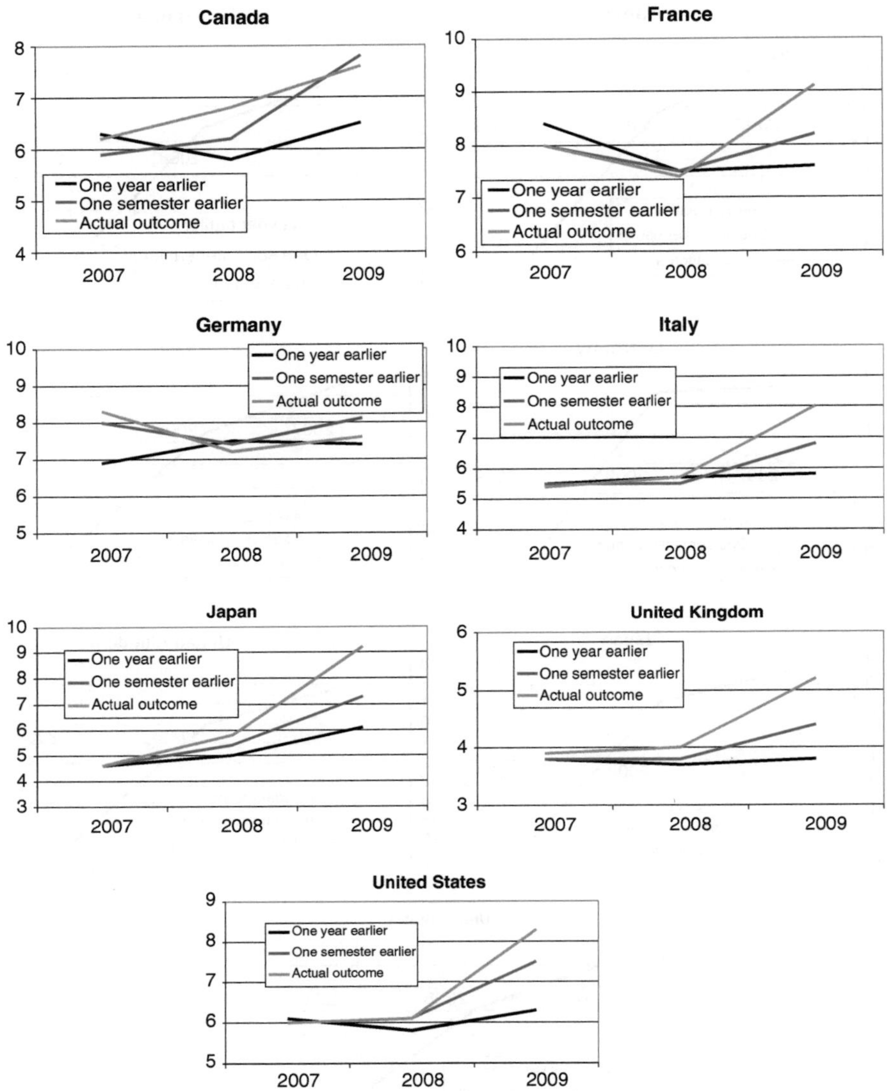

Fig. 5 Unemployment and projections rates (as % labour force) (Data Source: OECD Economic Outlook)

The pick up of exports in early 2009 was later reflected in a recovery of output, but not of private consumption, which continued to slide downward. Even the revival of growth in 2009 was not enough to close the widening shortfall of output gap compared to projection. In other words, output gaps should have been much narrower than actual values in 2009, indicating that either the policy measures were not adequate to bring output in line with potential, or a good portion of their impact was still to be felt, or both.

This interpretation fits all the G-7 countries, but with a varied timing: in UK and Canada private consumption turned up in the third quarter of 2009, while in Germany and France in the first. In Italy, the recovery of private consumption was also gradual and below projected trend. As to unemployment, government measures don't appear to have been effective enough in curbing the rise in unemployment rates, which was actually higher than projected in 2009, with the notable exception of Germany (a case that needs a deeper exploration).

On the whole, SMEs' output reflected this general trend, but with a more pronounced downturn in some sectors, such as manufacturing, because of their dependence on larger enterprises. In the financial domain, the impact of the measures aimed at easing credit access can be gauged at this stage on the basis of business surveys, as well as of the SMEs' use of the different government programs. Though restricted in scope,[40] the OECD survey of beneficiary SMEs offers some evidence about the effectiveness of government intervention in favour of SMEs finance. Firms see these measures as properly addressing the main constraints: 23% said the government programmes had helped them to survive and get more favourable credit terms, while 53% said they had helped save jobs. On the other hand, the majority of respondents did not believe the measures had been influential in obtaining better credit terms, and a third denied that they had been successful in opening up access to credit. So the evidence is rather mixed, but with a convergence of views on the need for better communication on these assistance programmes and improved management.

In the US, according to the NFIB Small Business Economic Trends surveys,[41] in August 2010 a net 12% of the responding SMEs found loans harder to obtain than 3 months earlier, representing a slight improvement with respect to the first quarter, while the percentage of regular borrowers was back to a record low of 31%. Likewise, the actual average interest rate paid by US small businesses on short-term loans was near the relatively high levels of 2009. In other words, financial conditions for US small business were not yet improving as expected.

In the Eurozone as well, banks' net tightening of loan supply was limited, but terms and conditions for SMEs remained tight and rather than improve actually worsened between March and September 2010.[42] The easing of banks' standards for granting credit since the fourth quarter of 2008 benefited large firms more than small ones; this was the case in Italy, where lending flows to SMEs actually turned negative at the end of 2009. Furthermore, interest rate spreads between small firms and large ones have risen, according to surveys by ISAE [43]

[40] Overall, only six countries were covered by this survey, with Italy the only country among major economies, while the other five countries had smaller economies, ranging from advanced ones, like Belgium and Finland, to emerging ones, such as Romania, Czech Republic, Estonia.

[41] National Federation of Independent Business.

[42] ECB.

[43] ISAE.

and Banca d'Italia. Lending stringency nevertheless eased, most likely thanks to the very accommodating monetary and financial policies. According to the estimates of Caivano et al.,[44] from 2008 through 2010 internal financial factors led to a cumulative net deviation of Italy's GDP growth of 0.5% below trend; but this negative impact was reduced by 0.7% points (from−1.2% to−0.5%) between 2009 and 2010.

In any event, there was no definitely established trend to easier credit conditions either in 2009 or in 2010. Signals coming from the ECB and NFIB surveys were mixed. This state of affairs was reflected in SMEs' heavy use of public credit guarantees, official credit and export credit.[45] Public guarantees on loans were the most used form of financial support, but went mainly to medium-sized firms, particularly those involved in external trade. Small firms were, instead, the main beneficiaries of other measures, such as credit mediation in France and Belgium, and the Mutual guarantee schemes in Italy. Guarantees were crucial in maintaining credit flows to SMEs as banks grew increasingly risk-averse and reluctant to lend. But as investment expenditure fell due to the recession, guarantees served mainly to fill gaps in working capital.

Certainly, one of the most effective support measures was the Government-sponsored agreement between banks and SME associations in Italy for the roll-over of maturing debt. About 257,000 SMEs applied for debt amounting to more than € 70 billion.[46]

Other forms of financial support, particularly for equity and quasi equity instruments, or innovation plans did not receive the same attention by firms, most likely because the recession delayed investment and financial restructuring, as evidenced by the decline in mergers and acquisitions. In Italy, this is also the consequence of the time taken by the authorities to make the new measures operational. It is, therefore, expected that the full impact will take longer to come through.

While export credit has been largely utilized and appreciated by firms, there is little evidence on the effects of structural improvements in SMEs' business environment. The same can be said for the firms rescued through public intervention.

Overall, macroeconomic support polices and public guarantees for SME borrowing helped to ease the recession and sustain the recovery of all enterprises, regardless of size. In this respect, small firms did not receive significant preferential treatment in the strategies of the G-7 countries, despite the countless governmental declarations of special concern for their plight. In the end, however, the real effectiveness of these policies should be evaluated in terms of structural improvements, for that is what will eventually make it less likely that SMEs will have to bear the brunt of financial crisis and recession, as they have in the past two or 3 years.

[44] Caivano et al.

[45] OECD e.

[46] At end-October 2010, 183,000 applications had been accepted , amounting to about € 55 billion of debt.

5 The Aftermath of the Recession

By the end of 2009, the economic recovery was already underway in all the OECD countries, but at different speeds, confronting governments with difficult choices on whether to continue support measures, especially for the SMEs. In the fast recovering economies, such as Germany, Canada and Switzerland, support was terminated by end-2010 and replaced by structural adjustments and selective assistance to some categories, like young innovative firms. Their emphasis actually shifted to promoting entrepreneurship, innovation, simplification of business requirements, internationalization, advisory services and financial access as a means to improve the business environment. For instance, Germany let the credit and guarantee programme expire at the end of 2010 as planned, but extended the special support for short-time work and pursued structural improvements.

Countries with sluggish recovery, such as the UK, France and Italy, instead, extended their assistance programs until 2011, though focusing on low cost measures which would not add to the already inflated fiscal deficits. For instance, Italy responded to the SMEs' growing demand of loan guarantees by increasing the scale of operations. European countries with unsustainable fiscal imbalances had, however, to reinforce their assistance programmes in spite of lack of resources, in order to prevent a fall back into recession.

In any event, all countries clearly perceive their support to SMEs as essential to expand employment, since these firms have been responsible for most of the job creation of the past decade. The US in particular launched a far reaching programme called "Startup America" to boost entrepreneurship and "The American Jobs Act" envisaging a easier access to capital for entrepreneurs. Across countries, there is also widespread recognition that capital and credit markets have not returned to a proper functioning and this market failure is penalizing particularly small firms, as shown by some indicators.

The OECD Scoreboard signals that the SME share of total bank lending is back to pre-recession levels only in a few countries, France and Italy among them, but not the US and other countries.[47] Mixed signals also come from the Euro Area Bank Lending Surveys.[48] Availability of bank loans was viewed as slightly improving in the second half of 2010, but interest rates and lending terms were worsening, especially for the riskiest borrowers. Banks' lending attitude is also likely to be affected by the mounting tide of non-performing loans. In any event, the stringency of credit supply is somewhat blurred by the weakening of economic activity which is reducing SMEs' demand for loans.

Equity financing was instead considered not relevant by 78% of surveyed SMEs, and actually, only 6% of them resorted to external sources of risk capital.

---

[47] OECD g.

[48] ECB a.

Overall, 2 years since the end of the recession SMEs are still in a precarious state, while governments' ability to continue assistance programmes appears more limited because of the fiscal retrenchment.

6 Unfinished Tasks and Conclusions

The experience to date with government polices for SMEs in response to the recession leaves a number of issues still in need of a durable solution.

The first issue is ways and means to make SMEs less vulnerable to changing conditions in bank lending. To expand public guarantees is little more than a stopgap that is necessary to tackle financial market failure to properly assess borrowers' creditworthiness. A permanent solution would rather require, on the one side, firms less unbalanced in their capital structures and more transparent in their behaviour, and on the other side, banks more attentive to financing small firms. As undercapitalization is at the core of SMEs' financing difficulties, more forceful policies would be needed to create an incentive for entrepreneurs to boost their equity base in lieu of resorting to excessive financial leverage. A key prerequisite here is the development of institutions specialized in risk capital investment, together with greater openness on the part of entrepreneurs to accept outside investors in their equity base. Private equity, business angels and quasi-equity instruments, such as mezzanine debt and participatory loans, need more support from governments.

Second, the relationship between banks and SMEs tends to be ruled by the automatic application of backward-looking scoring and rating methods to assess credit merit, with a view to optimizing bank capital absorption when deciding on a loan. The Basel two accord has further heightened this attitude, while marginalizing efforts to better know the ability of the entrepreneur and his managers to run a successful business. This approach could be harmful for SME financing in the aftermath of the deepest recession in more than half century, since it would imply attaching a heavy weight to the adverse effects of the recession on firms' balance sheets. The new Basel three rules do not squarely address this problem, but to a limited extent, as they recognize the importance of qualitative knowledge of local factors in credit assessment. Instead, it is essential to promote an approach to banking that revaluates the importance of a close relationship between the small entrepreneur and the lenders.

Third, in devising assistance programmes, small firms should be treated separately from medium-sized firms, because their needs differ significantly. It has become evident that they need specially designed support tools; otherwise they will be only marginal beneficiaries of government assistance measures. For instance, mutual credit guarantee schemes deserve special preference in government support programmes, because they better respond to small business financing needs, perform a measure of credit risk sharing among firms, and accordingly have an incentive for thoroughly screening and monitoring participating firms.

Simplification of tax and administrative requirements is another area where small firms need special treatment, since their thin organizational structure cannot cope with the red tape that normally accompanies business activity. Tutorship and management assistance are a necessary ingredient, since small firms' financial problems often stem from limited skills and knowledge of financing opportunities and financial management. Small firms need publicly funded assistance in terms of counselling and tutorship to upgrade management, improve finances and develop effective marketing approaches.

Fourth, small firms can grow best if they act in a functional relationship with other firms, particularly, larger companies. Being small and isolated is no recipe for success, because small firms can only reap the advantages of a larger scale by joining together with other firms. As we have seen, being part of a value chain can contribute to spread the problems of larger firms during a downturn, but at the same time it helps to establish a close cooperation, that is instrumental to a fairer sharing of the benefits and costs of boom and bust. Government policies, therefore, need to promote forms of aggregation more effectively than they have done so far.

Fifth, a key to strengthening a firm's competitive position during and after a recession lies with research and innovation, an activity that small firms, except those in technology-intensive sectors, find it hard to manage and fund. During the recession, most OECD countries stepped up their support to innovative SMEs that invest in R&D and innovation. But this did not prevent them from postponing or downscaling these programs. Thus, more effective Government measures are needed to help small firms to develop these activities through cooperation with research centres and other firms to carry out programs that go beyond their stand-alone ability to manage and fund over the business cycle. Innovation chains, extending even across the national borders, have proved a fruitful approach to bringing small firms into major projects and enhancing their ability to compete. It is also in this direction that Government support should be oriented.

Sixth, on international trade, Government support to SMEs during the recession consisted chiefly in expanding export credit and credit guarantee facilities. These were exploited mostly by medium-sized enterprises, that had already gained market share abroad, while small firms did not benefit as much, since they are less capable to enter foreign markets. Yet these markets are important for any firm to grow and become more competitive, as extending the market reach is a condition for any firm to expand. What small firms lack is the ability to know foreign market opportunities and to establish a market presence abroad. As a consequence, in addition to providing financing for foreign trade, Governments should support assistance services for small firms interested in boosting foreign sales.

A general shortcoming of policy response to the recession was the long time that authorities took to implement support programmes for SMEs, particularly of non-financial nature. This reflected the longer decision-making process that usually marks these measures by comparison with monetary and financial policies. It is likely, therefore, that the full impact of structural measures will become significant in the coming years, strengthening the currently hesitant recovery of SME business activity.

On the whole, these six areas for more effective public intervention are relevant in and of themselves, but all the more so at a time when governments are already discontinuing a number of assistance tools. As the exit strategy becomes a necessity for lack of resources, it is more urgent than ever that authorities shift the policy focus towards structural aspects, that can produce long lasting benefits for SMEs. The sense of urgency derives not only from the complexity of the solutions required by these problems, but from the risk that an attitude of constant dependency on special preferences and aid takes root among SMEs. Only decisive progress in these areas can reduce the likelihood that costly public aid is required on a quasi-permanent basis just to keep the SME sector alive and able to sustain the market challenge.

References

Banca d'Italia a, Relazione annuale, Anno 2009, 31 maggio 2010

Banca d'Italia b, Indagine sulle imprese industriali e dei servizi, anno di riferimento 2008, INVIND, Supplementi al Bollettino Statistico, luglio 2009

Banca d'Italia c, Indagine sulle imprese industriali e dei servizi – Anno di riferimento 2010, in Supplemento al Bollettino Statistico, Anno XXI, no. 37, 28 July 2011

Blinder AS, Zandi M How the great recession was brought to an end, 27 July 2010

Caivano M, Rodano L, Siviero S La trasmissione della crisi finanziaria globale all'economia italiana. Un'indagine contro fattuale, 2008–2010, Banca d'Italia, Questioni di Economia e Finanza no. 64, Aprile 2010

Codogno L Italy's economic outlook and the economic recovery plan, Ministero dell'Economia e Finanze, Dipartimento del Tesoro, 23 July 2009

De Socio A La situazione economico-finanziaria delle imprese italiane nel confronto internazionale, Banca d'Italia – Questioni di Economia e Finanza, no.66, Aprile 2010

EIM In: Audretsch D, van der Horst R, Kwaak T, Thurik Zoetermeer R (eds) First section of the annual report on EU small and medium-sized enterprises, 12 Jan 2009

EIM, IKEI EU SMEs and subcontracting, Final report, Oct 2009

European Central Bank a, The Euro area bank lending survey, Apr 2010–July 2011

European Central Bank b, Survey on the access to finance of SMEs in the Euro area: second half of 2009, 16 Feb 2010

European Commission a, European SMEs under pressure, annual report on EU small and medium-sized enterprises 2009, DGEI, released on 11 May 2010

European Commission b, Progress report on the implementation of the European economic recovery plan, Dec 2009

European Commission c, The European recovery plan

European Commission d, Report on the implementation of the SBA, COM(2009) 680, Dec 2009

European Commission e, Seventh report on potentially trade restrictive measures, Sept 2010

Hempell HS, Sorensen CK The impact of supply constraints on bank lending in the Euro area crisis induced crunching? ECB working paper series no. 1262, Nov 2010

ISAE Rapporto Isae, Le previsioni per l'economia italiana, luglio 2008

Mediobanca Indagine sulle medie imprese, 2010

National Federation of Independent Business In: Dunkelberg WC, WadeH (eds) NFIB Small business economic trends, Apr and Sept 2010

OECD a, Turin round table on the impact of the global crisis on SME and entrepreneurship financing and policy responses: issues paper, 26–27 Mar 2009

OECD b, WPSMEE, Assessment of government support programmes for SMEs and entrepreneurs' access to finance in the global crisis: monitoring report, Mar 2010

OECD c, WPSMEE, Inquiry to beneficiary SMEs survey: results and analysis, Sept 2010

OECD d, WPSMEE, Assessment of government programmes for SMEs and entrepreneurs' access to finance in the global crisis: updated monitoring report, Sept 2010

OECD e, Economic outlook, no. 84–87, 2008–2010

OECD f, Working party on SMEs and entrepreneurship, A pilot OECD Scoreboard on SME and entrepreneurship financing data and policies, 17 Nov 2010

OECD g, WPSMEE Steering group, An OECD Scoreboard on SME and entrepreneurship financing: preliminary data and analysis, working document 15–16 Sept 2011

Panetta F, Signoretti FM, Domanda e offerta di credito in Italia durante la crisi finanziaria, in Banca d'Italia, Questioni di Economia e Finanza, no. 63, Aprile 2010

Zecchini S The impact of the global crisis on SMEs and entrepreneurship financing, in OECD and Intesa San Paolo, The impact of the global crisis on SME and entrepreneurship financing and policy responses, 2009

Industrial Policy for SMEs Renewal: The Opportunity of Service Platforms

Bruno Basalisco and Guido M. Rey

1 Introduction

Advances in network technologies and related services can be a key factor supporting firms' innovation and growth. In particular, the networked enterprise business model and the underlying enabling technologies offer a vast potential to firms willing to invest in this innovation. The adoption of these technologies varies across industries and firms' sizes, which we will further elaborate on the basis of the survey data analysed in what follows. As a consequence, economies where SMEs play a major contribution and where this is confined mainly to low-tech traditional industries, show levels of diffusion of network technologies below international averages. Italian SMEs, with respect of the process of ICT diffusion, fail to realise the expected complementarities between ICT and the associated investments in human capital and organisational change.

On a similar note, the delay across the whole of Italy in embracing the technical and organisational change associated to ICTs threatens to foster a split between Italian businesses and their counterparts in the global economy: Italian SMEs risk being trapped in a business digital divide. Today, digital, sectorial and geographic divides are intertwined and compounded; in a "flat" world where competitiveness is informed by the breadth and depth of each country's connectivity (Friedman 2006), these divides threaten to weaken the performance of the Italian economy.

This version: November 2011

B. Basalisco (✉)
Istituto di Management, Scuola Superiore Sant'Anna, Piazza Martiri della Libertá, 24-56127 Pisa, Italy
e-mail: brunix@yahoo.com

G.M. Rey
Istituto di Management, Scuola Superiore Sant'Anna, Piazza Martiri della Libertà,
24-56127 Pisa, Italy
e-mail: guidom.rey@gmail.com; g.rey@sssup.it

G. Calcagnini and I. Favaretto (eds.), *Small Businesses in the Aftermath of the Crisis*,
Contributions to Economics, DOI 10.1007/978-3-7908-2852-8_3,

Many firms have deployed ICT while switching to a new business model. Scholars who have focused on corporate investment on ICT broadly agree that, for corporations to maximise the benefit from using advanced ICT infrastructure, significant business transformations must occur.[1]

The extant literature mainly explains what makes a firm's ICT investment successful (or not), which can facilitate and thus encourage ICT deployment strategies. Nonetheless many firms currently fall short from tapping into the benefits of advanced ICT infrastructures and services, which would be beneficial to society as a whole. Thus a research gap exists as to the normative issue of how to make sure that those laggard firms are instead able to fruitfully invest in ICT. We believe that such line of research is a valid response to the key observation made by Bresnahan and Trajtenberg, who showed that, because of its general purpose technology (GPT) nature, the development and match of supply and demand of ICT present peculiar challenges: "in particular, if the relationship between the GPT and its users is limited to arms-length market transactions, there will be 'too little, too late' innovation in both the GPT and the application sectors" (Bresnahan and Trajtenberg 1995, p. 103). A policy aimed at ameliorating such technology-dependent market failure is therefore a relevant research subject.

If a given firm's ICT investment can be more successful when matched by the move to a revised business model, we argue that this intuition can be transposed within the policy – rather than corporate – domain. We do so by introducing a policy equivalent of the business model construct, which we will refer to as "policy model", specifically a platform industrial policy. We embrace this novel approach as we believe that it enables both a fruitful positive analysis of the systemic challenges that prevent a silent majority of firms (frequently the smallest ones) to benefit from advanced ICT – as well as normative suggestions as to a potential avenue for a policy of industrial development.

The platform policy model is in fact seen here as a tool to devise policies aimed at mitigating the structural barriers to the adoption of novel adaptive infrastructural technologies and business models by SMEs. In order to do so, we will draw from the insights of the burgeoning literature on platforms in the fields of industrial

[1] There is a wealth of literature in economics and management strategy analyzing how firms can stand to benefit from investment in their ICT systems. An oft-quoted example is the clothier Zara, whose business model is highly synergic to the ICT infrastructure embedded throughout its supply chain (Ferdows et al. 2004) – providing the firm with a strong competitive advantage. Some authors stress that in order to yield benefit to a firm, ICT investment needs to be matched by some changes in its organization and business. For instance, a successful organizational adoption of ICT is weaved with the management of knowledge processes (Antonelli 1997a, b). If the latter does not match the former, the end results on firm productivity of the new technology may be mitigated or lost completely. This phenomenon can be generalized by stating that for firms there are complementarities in production between internal reorganization and ICT investment. Finally, taking a techno-economic perspective, it has been argued that ICT follow in the footsteps of the steam engine and the electric motor in embodying the general purpose technology (GPT) archetype, because of their extensive breadth of applications and impact on economic growth (Bresnahan and Trajtenberg 1995).

organisation and strategic management. This allows us to highlight why and how it is beneficial to induce SMEs to join services platforms; we will stress the pivotal role that market-compatible "nudge" policies can play in the renewal of SMEs.

2 The Strategic Links Between the SME Industrial Context and the Policy Approach Analysed

The technology adoption story in the economics and business literature is frequently centred on the incentives informing the decision of a single firm (as to whether to invest in a new technology or not), with respect not only to its competitors' actions but specifically when the technology supports complementarity between products and services (downstream or upstream), and when compatibility issues affect the strategic incentives to innovate. More efficient outcomes can be associated to the level of coordination between firms and this, in turn, boils down to trust and to the institutional framework underlying firms' access to markets.

While network effects can drive fast adoption of successful and popular ICT consumer technologies and services, these adoption patterns are not always the case: as well as many take up stories, so have several unsuccessful network products or services entered and exited the markets. This follows from poor coordination in joining networks: an important challenge for Italian SMEs to overcome.

First, the higher the number of actors (consumers, firms or NGOs such as standards bodies) composing a sector is, the more difficult it is for sufficient coordination to arise. This can be associated to the free-riding problem (analysed in game theory): where there are spill-overs from investment (e.g. in network technologies), the incentives for some agents to exploit other agents' investments (without any contribution) increase as the number of members of a community increases. As a result, every time a single firm goes alone in investing in a network technology, other firms may subsequently benefit from the risk reduction and perhaps from connecting to the deployed infrastructure.

Second, on top of the above, an important question that is clear in the mind of every business assessing the benefit of networking technologies is whom can these technologies allow to meet and whether existing and potential clients can be found on the network. Therefore, it is not only the number of participants on a network but also their kind that will matter in adoption decisions and in business investment. This suggests that a platform – rather than simply a network framework – can best represent the issues surrounding business use of ICT products and services. In the light of this coordination challenge, we see that each firm's decision to invest in network technologies and in the organisational change required to leverage them will depend on the counterparts this business faces and on any institutional/regulatory characteristics which may mitigate (or exacerbate) the investment coordination problem.

In countries such as Italy, where a multitude of SMEs compose the backbone of the economy, the presence of both issues (free riding; counterpart uncertainty) makes ICT investment coordination extremely complex and obstructs the network adoption dynamics, thus explaining the lag in the diffusion of ICT services.

Therefore, there are structural reasons which call for special awareness by Italian policy-makers of the innovation challenge which the networked enterprise business model and technologies pose to Italian SMEs and as a consequence to the whole Italian economy.

The past successes of Italian SMEs have often been associated to an ability to be on the cutting edge of the low-cost dimension and an entrepreneurial drive to seek new markets. This business model is now mimicked by several emerging countries, which benefit from lower costs of labour and equivalent access to an ever more interconnected global economy. The low-cost concept permeates all cultural barriers and is aptly conveyed via the communication infrastructures (as an inspection of the eBay online marketplace will confirm for the C2C and B2C domains). Notwithstanding their legacy of low-cost production, Italian SMEs are currently striving to transition to a model based on competitive advantage derived from high-quality production, which is less replicable and therefore more sustainable. The journey which must be undertaken today by a firm wishing to sell the quality of its products is significantly different from the venturing salesman journey which may have been sufficient to market low-cost products a few decades ago: quality is intrinsically intertwined with culture, which implies that intangible resources and service provision are key future drivers for Italian SMEs to succeed at marketing quality manufactured goods.

The quality strategy requires a host of complementary services to support not only the production process per se (and the underlying organisation) but also the ability for firms to market their quality goods effectively on a global scale. This is because in a business world where the value paradigm is defined by the services content (even for manufactured goods) the factory – while an important source of value adding activities – is not anymore the central component.

In order to leverage ICT for the purpose of adding value to production activities, a fast internet connection (which in any case is not to be taken for granted in many areas of Italy) and access to eBay are not simply enough – as many SMEs are well aware of.

As stressed by marketing scholars, a new service paradigm is now the perspective through which all economic transactions – including those pertaining to manufacturing activities – must be seen in order to understand their value (Vargo and Lusch 2004). Because of this, a key challenge for manufacturers is the overcoming of cultural barriers between supplier and customer; this is harder for SMEs to perform without the support of dedicated advanced professional services which at the moment only large companies find available and can afford to tap into.

3 A Survey Analysis of Italian SMEs

In the global marketplace and economic system, small businesses and traditional local production systems have difficulty in maintaining their strategic positions as in this new scenario firms' reliance on flexibility is not viable today, neither from an

Table 1 Turnover and persons employed per class of person employed and branch, year: 2009

Branches NACE classes of person employed	Turnover	Persons employed	Value added per person employed	Labour cost per employee
Economic activities	(Comp. %)		(Thousands €)	(Thousands €)
Industry (C,D,E)				
1–9	10.1	23.4	25.3	25.4
10–19	9.2	14.4	36.6	29.1
20–49	12.9	15.7	44.7	33.5
50–249	26.0	21.6	54.9	39.3
250-and over	41.8	24.8	73.0	46.4
Total	**100.0**	**100.0**	**48.2**	**36.8**
Services (G,H,I)				
1–9	34.0	54.7	24.0	24.1
10–19	11.8	9.0	36.5	28.8
20–49	11.2	7.3	42.6	31.9
50–249	17.1	9.7	41.4	34.0
250-and over	25.9	19.2	52.1	35.3
Total	**100.0**	**100.0**	**33.6**	**30.7**

analytical nor from an economic policy perspective investments in intangibles resources have achieved new relevance.

Innovation requires applied research and the ability to exploit it and should also involve marketing, services and organization (Chesbrough 2011). This innovative strategy needs a favourable environment, but Italy has a multitude of micro and small companies that rarely innovate, few medium-sized enterprises and very few large firms.

The recovery from the present crisis depends, amongst other things, on structural soundness and conditions of profitability and competitiveness and both are related to labour productivity and wages.

From Table 1, comparing productivity and wages of the first class with the same items of the last class it can be seen that productivity of the latter is three times that of the first class while in comparing wages the difference is only twice.

Comparing the outcome of the 2 years 2008–2009, all classes suffered diminishing revenue and decrease in productivity (−4.1% on the average for industry and −7.8% for services) but for wages the decrease was only −0.8 and −1.8% respectively.

From analyses conducted on the impact of the crisis, indications can be drawn that progressive companies are emerging acquiring resources, competencies and capabilities to find and exploit new opportunities. These medium-sized enterprises have or could have a leading role in the Italian industrial districts since they are able to impose network services to reduce the time of the production process and the inventory levels, to improve logistics and to increase exports and imports substitution.

Medium-sized enterprises are certainly innovative by vocation and due to the need to grow - if not for their survival – and with their network of relationships could create the conditions for technology transfer and dissemination of network services in micro and small enterprises, particularly in either industrial or business districts. Manufacturing activity is largely delegated to networks of small businesses, firms that already produce innovative components at lower costs (sometimes for several competitors). By means of these innovations, medium sized enterprises provide advancements, on the industrial system level, due to their interest and ability to take the lead in firm networks, enabling improvements in the efficiency and communication of know-how, as well as stimulating change in their perimeter of their suppliers, thus giving rise to a particular inter-organisational model with its own governance and economic logic (Rey and Varaldo 2011 p.767).

These are open operational models that should be carefully looked at, as a lever to extend the application fields of ICT and for the dissemination of network services in the small business sector, overcoming their resistance to digital innovation (Rey 2011).

The wealth of knowledge of an organization includes not only the intellectual capital and know-how of individual employees but also that of their firm (the "corporate culture") and the correlated supplier system. In evolutionary terms, supply and subcontracting networks can be developed and transformed from "supply chains" into "value chains", where the contribution of the various components becomes crucial and irreplaceable for innovation partnerships with shared values, benefits and risks.

Surmounting the traditional geocentric logic of industrial clusters, still valid but with increasing limitations, could occur through giving economic-institutional value to networks that are taking shape as a type of "business district", reproducing many of the traits and values of dynamic entrepreneurship of the systems of Italian small artisan-style production units (Rey and Varaldo 2011, p.772).

3.1 The Diffusion of ICT in Italian Enterprises

The innovations induced by ICT and in particular by the network services have not brought about the expected changes in processes, products and organisation, although one can read a positive sign in the slow bridging of the gap vis-à-vis our European partners.[2]

In the first 10 years of the twenty-first century, network services provided by ICT have spread in Italy but the data suggest that Italian firms have given a cool reception to this opportunity and in general these services are utilized to merely exchange information. Technologies are used to reduce direct and indirect administrative costs but enterprises are reluctant to use network services to reduce transaction costs, to improve their market share and to supply services connected

[2] A comparison with other EU countries places Italy in an intermediate position overall but lagging behind when compared with countries of similar size and level of development Giannakouris and Smihly (2010), Tables 7.19 and 7.20).

Table 2 Enterprises with at least ten employees and with Internet access, by type of online service used, 2010[a] *(percentage values on the total of enterprises with Internet access)*

Business activity / Number of employees	Access to banking and financial services	Acquisition of market data (e.g. prices)	Acquisition of services and data in digital format	Acquisition of post-sale services	Staff training
Business activity					
Manufacturing	**90.2**	**69.4**	**56.5**	**55.6**	**18.6**
Total services	**83.4**	**59.9**	**53.7**	**47.4**	**25.7**
Information activities	95.0	79.0	83.9	66.9	58.7
Size class					
10–49	85.5	64.0	53.0	50.1	21.1
50–99	94.0	78.9	72.7	58.8	30.1
100–249	97.1	74.8	74.1	54.7	37.9
250 +	94.8	84.8	83.8	61.9	50.5
Italia	**86.6**	**65.5**	**55.3**	**50.9**	**22.6**

[a]Several answers to the same question are possible on the questionnaire. So, the sum of the percentages per line may be over 100

with the transaction. In summary, technologies are not intended to substantially modify the organisation of the firm or its governance.

It is difficult to identify the contribution of ICT to the growth of value added but there is no doubt that network services have not been able to counter the deflationary drift accompanying the decline in competitiveness of the Italian ecosystem. Unfortunately, in the absence of an industrial policy that promotes the diffusion of network services, firms are cautious and wait for clear signals from competitor and supplier behaviour and in some respects from households (Acconcial et al. 2009). From the ISTAT survey carried out in 2011 on the diffusion of technologies for network services (ISTAT 2011), it emerges that the most utilised services by the enterprises are banking services, through the inter-banking network (see Table 2).

Minor differences are due to the business activity and the size class, except for staff training, the least utilised, but it is interesting to note that services enterprises are in a better position with respect to manufacturing and that the smallest class is not even the half of the last class.

About two thirds of the respondents has a web site or a home page to deliver services (Table 3); this share increases with the size of the firms but there are relevant differences following the diverse services offered on internet, both in business activity and size of the firms.

It is relevant to note that a comparison with the 2008 survey shows that the percentages in most cases have nearly doubled in these 3 years.

The analysis of the data on the network services deserves more careful consideration to detect their possibility to improve processes and to reduce transaction costs, especially between enterprises (Table 4). Approximately two thirds of manufacturing firms and less than half of total services firms transmit electronically information to/from other enterprises suitable for automatic processing of data.

Table 3 Enterprises with at least ten employees and with a Web site or home page, by type of online service offered, 2010[a] *(percentage values on the total of enterprises)*

Business activity/ number of employees	Have a web site or a home page	Services						
		Product catalogues or price lists	Online ordering, reservation or booking	Advertisement of open job positions or online job applications	Possibilities for visitors to customise or design products	A privacy policy statement, privacy seal or certification related to website safety	Order tracking available on line	Personalised content in the website for regular/repeated visitors
Business activity								
Manufacturing	**69.7**	**48.8**	**11.4**	**6.4**	**4.5**	**32.6**	**4.3**	**2.0**
Total services	**60.4**	**35.8**	**18.6**	**9.4**	**5.0**	**30.1**	**7.3**	**3.5**
Information activities	80.7	43.4	13.0	30.5	5.0	44.9	4.7	3.7
Size class								
0–49	58.8	36.0	13.2	5.5	4.1	27.3	4.7	2.4
50–99	82.0	52.1	15.1	16.5	4.7	43.2	8.5	3.4
100–249	81.8	49.5	17.8	25.5	5.4	48.5	8.4	2.8
250 +	89.9	53.1	22.8	47.8	7.7	62.8	14.5	7.0
Italia	**61.3**	**37.6**	**13.6**	**7.5**	**4.2**	**29.5**	**5.2**	**2.5**

[a]Several answers to the same question are possible on the questionnaire. So, the sum of the percentages per line may be over 100

Table 4 Enterprises with at the least ten employees transmit electronically information to/from other enterprises suitable for automatic processing of data, 2010[a] *(percentage values on the total of enterprises)*

Business activity/ number of employees	Electronic transmission of data	Used for					
		Sending order to suppliers	Receiving a e-invoice	Receiving orders from customers	Sending a e-invoice	Sending or receiving product information	Sending or receiving transport documents
Business activity							
Manufacturing	**67.8**	**52.3**	**60.1**	**55.5**	**40.3**	**55.0**	**41.0**
Total services	**60.5**	**44.7**	**49.7**	**36.9**	**26.4**	**46.6**	**28.4**
Information activities	76.6	55.5	63.6	49.3	44.2	50.2	37.1
Size class							
0–49	62.1	47.5	54.3	43.3	31.0	50.4	31.8
50–99	72.0	52.5	58.5	50.1	39.7	55.4	41.8
100–249	71.5	47.8	51.8	48.8	38.3	49.0	40.1
250 +	75.5	48.8	49.7	48.8	43.8	48.5	44.3
Italia	**63.2**	**47.8**	**54.4**	**44.0**	**31.9**	**50.6**	**32.9**

[a]Several answers to the same question are possible on the questionnaire. So, the sum of the percentages per line may be over 100

Table 5 Enterprises with at least ten employees share electronically information with customers and suppliers on inventory levels, production plans, demand forecasts and share deliveries, 2010 *(percentage values on the total of enterprises)*

Share electronically information with suppliers and customers	Inventory levels, production plans, demand forecasts and deliveries	Suppliers	Customers	Instruments	
				Use a web site	Use EDI, XLM, EDIFACT
Business activity					
Manufacturing	**21.7**	**17.8**	**19.9**	**15.4**	**10.7**
Total services	**23.4**	**19.9**	**15.2**	**18.4**	**11.8**
Information activities	37.5	31.3	30.8	30.1	21.5
Size class					
0–49	20.9	18.0	15.7	15.6	10.0
50–99	25.2	19.9	20.6	20.7	13.6
100–249	33.0	24.8	27.5	25.0	18.8
250 +	40.4	32.7	32.8	32.5	28.9
Italia	**21.8**	**18.5**	**16.6**	**16.4**	**10.8**

The highest percentage concerns received and sent orders, received invoices, transport documents, note that only one fourth of services firms send invoices electronically but it must be considered that the service sector has a larger share of families as customers. Two remarkable considerations are the limited difference due to the size of firms and the impressive increase of this network activity in the period 2008–2010, during which it doubled.

An even more interesting result concerns the share of information on inventory levels, production plans, demand forecasts and share deliveries between suppliers and customers (Table 5).

This activity involves only one fifth of firms without distinction between industry and services but differences in size amount to over 20 points. In the case of bilateral network relations, the aim is to obtain cuts in transaction costs, but also to improve the closing of orders and to monitor production timetables and/or logistics.

As to the diffusion of network service within the enterprise the situation have large possibility of increase (Table 6) because in 2010 only one fifth of the firms were utilizing this tools to ameliorate the efficiency and the velocity of administrative processes and it applies both to sales and purchases. The level of attention paid to inventory information must also be noted. As expected these percentages increase with the size of the firm and with the business activity as the largest occurrences are for manufacturing.

The puzzling outcomes of these surveys on the diffusion of ICT servicing concerns the e-commerce and the results show the increasing diffusion of the electronic purchases: 35.9% of the firms with a five points difference favourable to service firms but only 5% of the firms used any computer networks for sales and while there was an increase of about five points for purchases in the period 2007–2009, for electronic sales there was practically no increase. Both for purchases and sales, firms of 250+ employees show a greater diffusion than the average (59.1 and 16.6 respectively).

Table 6 Enterprises with at least ten employees share electronically information within the enterprise, 2010 *(percentage values on the total of enterprises)*

Share electronically information	On sales with the software used for any internal function				On purchases		Have ERP software package	Have CRM (software for managing information about clients)
	With the management of inventory	With the accounting	With the production or services management	With the distribution management	With the management of inventory	With the accounting		
Business activity								
Manufacturing	**26.6**	**40.9**	**31.6**	**22.4**	**25.9**	**36.4**	**27.7**	**24.0**
Total services	**19.8**	**31.2**	**24.2**	**18.6**	**19.5**	**27.3**	**19.8**	**25.2**
Information activities	18.9	43.2	38.8	19.6	18.9	37.7	35.1	41.4
Size class								
0–49	18.1	32.0	23.1	16.1	17.8	28.0	18.4	21.8
50–99	38.5	47.4	41.1	31.7	36.2	44.3	44.4	33.5
100–249	40.9	56.8	52.0	38.6	42.4	53.7	56.8	36.9
250 +	49.9	63.3	58.1	48.5	52.0	64.6	68.0	45.8
Italia	**20.5**	**34.2**	**25.6**	**18.2**	**20.2**	**30.3**	**21.9**	**23.4**

It is interesting to note that the manufacturing firms most active in the new technologies are those involved in import and export and this confirms the theory that international trade is a powerful stimulus to the introduction of innovative technologies not only in processes and products but also in business organisation and management. It also confirms that in the absence of an external stimulus, the diffusion of network services is slow and not always appreciated.

The surveys also show that once firms use ICT extensively, this will involve both multilateral relations and exchange of information and services, inside and outside the enterprise.

In particular, services provided by the network allow breakdown and delocalisation of the various production activities (e.g. logistics, administration, finance, archiving, etc.), increase in the efficiency of the production processes and widening of the actual and potential market, the latter option not widely known by the majority of the entrepreneurs.

The outcome of the 2011 ISTAT survey show that the strategy suggested in this article is already spreading in the Italian economy and it is expected to further increase in the innovative enterprises while conservative one are predestined to disappear. On-line services and the existence of an enterprises network is a strategic element integrated in the industry output. In short to improve the value chain, it is essential for enterprises to in-source those services that can foster integration between manufactured goods and services. This is the true meaning we should attribute to the motto *the factory is obsolete* because services cannot be limited to simple intermediation: they must become strategically part and parcel of the turnover of industry.

4 Alliances and Inter-firm Collaborations

During the last two decades, the study of networks of firm became a central topic in industrial economics and organizational studies from both a theoretical and an empirical perspective. In this respect there was a significant increase in collaborations and alliances among independent firms to share the risk of large investments in research, innovation and in the establishment of a wide range of networks (laboratory, technology, market, logistic, etc.) (Sallusti 2010).

Increasing was, also, the heterogeneity of organizational and relational structures that shape alliances and inter-firm collaborations. Governance structure forms are ranging from unstructured collaborations, through consortia and joint ventures that superimpose new governance structures on existing firms, to transactions that devise new firm boundaries and asset ownership. The difficulties with alliances are the associate acceptance of some obligations and the credibility of the pledge on their future behaviour and furthermore the importance to manage rightly the strength and weakness of the participating firms.

In industrial economics, cooperative relationships are generally considered as hybrid forms of governance, that is an intermediate organizational device between market and hierarchy, which are polar opposites.

There should be a virtuous equilibrium between sharing resources and knowledge vs. competition as the distinctive feature and the main rationale for network formation. Indeed, collaboration in managing interdependencies and some degree of competition give to efficiency of networks a plain advantage over other kind of organization.

A debate exists on the perspective from which inter-firm relationships (and, thereby networks) should be analysed between Transaction Cost Economics and Relational Contracting approaches on one hand, and Resource-Based and Knowledge-Based views on the other hand. The selection should define the relative efficiency of the different forms of governance, in order to determine the relative efficiency of both the organizational forms and the ability to manage interdependencies.

The former privileges the analysis of the inefficiencies caused by the presence of informative asymmetries. The latter, instead, privileges an analysis focused on the tangibles and intangibles resources involved in the transaction.

Criticisms about the completeness of transaction cost and incentive alignment approaches would involve a reduced possibility to analyse the influence of technological and functional complementarities and interdependence on the choice of governance. Instead, taking assets and transactions as the core of the analysis prevents acknowledging the role of learning in grounding the development of resources and alliances themselves and the relation that links governance forms and learning processes.

Notably, transaction cost economics and relational contracting view tend to characterise alliances based on their contractual structure, acknowledging the role of asset specificity, incentives and informative asymmetries, as the determinants of alliance design and governance structure. Alternative approaches tend, instead, to characterise alliances based on their organizational structure, acknowledging the role of resource characteristics in shaping alliance design and coordination strategy.

The innovative strategy we are suggesting needs a corporate renewal and a favorable environment to enhance alliances and inter-firm collaboration. Step of this evolution are the increase in firm size, the renewal of the industrial districts no more simply geocentric, the selection of firms for their ability to supply the commodities and services required by Italian and foreign market. This means that industrial districts should develop into business districts where the distribution of benefits and risks to the various components takes into account their contribution to the creation of the value added of the business. This aspect of the governance of the network organizations becomes crucial and irreplaceable for innovation partnerships. The difficulty is the selection of different shape of this collaboration that presently privileges the hierarchy and the market power, while the selection of the partners is based both on resource and knowledge in particular for the high quality of the Italian artisans.

At the present time, medium-sized enterprises (ME) have a leading role in most successful districts since they could impose network services to the small enterprises. With their network of relationships, (ME) lead technology transfer and dissemination of network services in foster improvements in the efficiency and communication of know-how and stimulate, as well, they would change their partners if they don't accept to manufacture innovative components at lower costs and if they resist the progress suggested by the leader ME (Rey and Varaldo 2011, pp. 769–771).

To convince SME to accept this renewal in the organization of the Italian industries it would be sufficient to recall that the present evolution of the Italian manufacturing suggest a dramatic reduction of firms and person employed. SMEs need boosting their position in the foreign markets, especially in the emerging markets, but this strategy requires a dramatic increase in their human, technological, financial and knowledge resources.

More difficult for them is the acceptance of this particular inter-organizational and/or inter-firms collaboration model with its own governance and strategy and the redistribution of powers according to these new forms of organizations. There is no more an external objective hierarchical power called market that judges the relative value creation of the single participant to the transaction and defines the governance and the distribution of benefits and costs within the members of the network.

5 An Alternative Approach to Platforms: Acknowledging Different Platform Concepts

The recent academic literature provides us with three distinct, though interrelated, approaches to understanding the functioning of platforms. These strands conceptualise similar phenomena while focusing on different issues and providing complementary perspectives on platforms:

- The economic perspective of two-sided or multi-sided markets (within the field of theoretical Industrial Organisation) studies the relationship between prices and adoption of the platform by the different groups of actors that interact on the platform, while generally taking as a given the existence of the platform itself (static model with respect to innovation). Platforms are seen as intermediary entities which enable and mediate transactions between two groups of agents. Platforms extract value inasmuch as they can support beneficial transaction (which will depend on existing externalities, i.e. the partaking of information) and can measure (and keep within the platform) the interaction between those on the platforms. The focus here is on the optimal level and structure of prices levied by the platform (to each class of users) as a function of the externalities (adoption and usage externalities across groups). We will refer to this viewpoint as **Platform A – Economics perspective**.
- The perspective within the complex systems management and management of high-tech industries literature (generally within the field of strategic management) considers a platform as an instrument for a firm to coordinate the

development of technological innovation based on standards. This viewpoint focuses on the platform firm's strategies to promote the adoption of its innovation within a highly interconnected industrial ecosystem (typically within the ICT industries). According to this view, platforms can facilitate innovation in technology standards by mediating and coordinating the interests of consumers and suppliers. We define this as the **perspective B – the platform as a standard**.

- The final viewpoint transcends a specific industrial category and is instead applicable across any scenario where a key firm holds some competitive advantage (defined for instance by patents or industrial secrets or characteristic skills in design or other fields where skills are appropriable) and leverages such a competitive advantage by opening its supply chain, transforming it into a platform. What this entails is for the key (i.e. pivot) firm to facilitate the intermediation between its suppliers (i.e. turning these into "complementors" in platform jargon) and its customers – up to the point where its suppliers can add value to the firm's product/s with a degree of independence (i.e. with a business model which is more diversified and sustainable independently from the custom of the key firm). The reason why a firm would want to open its supply chain is to enhance its potential market by cajoling (existing and new) suppliers to enlarge the range of complementary features of the key product. In turn this can encourage its clients to enlarge the range of solutions, services and goods which incorporate the key product – thus increasing its appeal and opening potential new markets. We will refer to this perspective as **Platform C – the platform as an open supply chain**.

Note though that more than one of these aspects may coexist within the same platform, which explains the current ambiguity in the academic conceptualisation of platform; as well as explaining the strong and sustained interest in the platform construct across disciplinary boundaries. For the reasons outlined above, this work transposes the use of the platform construct onto the public policy domain; specifically we argue that, because of their multiple facets, platforms can enable a successful pursuit of public policies aimed at connecting and empowering marginalised economic actors by leveraging their diverse interests, characteristics and capabilities. All three perspectives are useful to the policy-maker to understand the evolution of competition and interdependence in a "flatter" world. At the same time, given the urgency to provide pragmatic policy advice (for policy-makers to intervene smartly by "nudging", or altering subtly but effectively firms' incentives), we will focus our analysis on applying the viewpoint C, since it is more adaptable to different industrial contexts and captures that of SMEs intensive districts and industries such as those underpinning the Italian economy. After reviewing the key learning points from the academic literature, we will thus propose two possible platform models, informed by perspective C; these models specifically aim to emancipate SMEs, by leveraging (as an asset rather than liability) their role in the supply chains that characterise their district/filiere industrial context.

5.1 *Complementary Perspectives on Platforms in the Academic Literature*

A recent flurry of economic literature discusses the incentives of a platform – firm which can earn revenues from more than one side of the market. Two-sided markets are those where platforms revenues – coming from two distinct groups of actors – depend on the interactions between the different categories of subscribers to the platform (Armstrong 2006; Hagiu 2006; Rochet and Tirole 2003, 2006; Weyl 2009). A platform firm succeeds inasmuch as it can generate value by enabling transactions between users of the platform – and capture part of this value. These benefits depend of the nature of the externalities that the platform is able to internalise. When both sides of the market gain from the interaction with the other side, the platform firm faces a dynamic challenge commonly referred to as the chicken-and-egg problem (Caillaud and Jullien 2003). This issue occurs in two-sided platforms exactly because they differ from merchants (Hagiu 2007). A new merchant could enter a market by investing part of its capital on stocking supplies; having done that, it could then proceed to resale at a marked-up price. Instead platforms do not "stock up" one side of the market: both sides of the market have to interact simultaneously.[3]

On the other hand, inasmuch as platform industries present remarkable rates and patterns of innovation, evidence of market creation (and of the strategic efforts underlying such creation) suggests that the assumption of a given set of markets is untenable. A key driver of business innovation is in fact firms' – whether large or small – aims to expand beyond a given market type.

Pricing is only one side of the innovation story in platform businesses. Firms such as Intel which aim to increase the systemic value of a general purpose technology such as the microprocessor, use a variety of non-price strategies to induce value-enhancing complementary innovation (Gawer and Cusumano 2002). Platform owners who strive to orchestrate innovation around a proprietary innovative core (Platform B) do not confine themselves to promoting adoption decisions in existing markets which depend on the platform. They seek to engage complementors (firms supplying complementary products or services) into developing new markets – adopting both pricing and non-pricing strategies (Gawer and Henderson 2007).

[3] Moreover, unlike merchants, platforms take minimal liability for any goods/services provided by one of the sides to the other and intermediated by the platform itself. This is currently the case in many virtual marketplaces (Platform A), although some actors such as eBay/PayPal are introducing procedural safeguards to reassure bidders against potential frauds. Instead, online portals (frequently based on a "walled garden" model) are expected to guarantee the content of the goods and services associated to their environment. This is the case in many mobile internet portals which enable their users to purchase and download applications such as ringtones, etc.

5.2 Complementors and Market Creation

Industrial ecosystems present considerable complexity than what reflected in any if the strands of the platforms literature: SMEs clusters/districts are a prominent example of forms of industrial organisation where complementarities in products and services play a key role (supported by local proximity). The platform literature so far recognises that in key high-technology industries some firms play a pivotal role in orchestrating innovation throughout their ecosystem – both with downstream and upstream partners. While the former concerns adoption of innovative technologies in user markets (which is fundamental for the underlying platform product to hold a competitive edge), the latter revolves around enhancing the platform value by adding complementary components specifically honed to the platform characteristics (Platform B). Hence, it has an incentive to promote product development in complementors. A platform owner attempting a type C strategy is unlikely to succeed in this pursuit, though, if it sets a one-off transaction as a "procurement" of the complementary products/services development sought. Major ICT firms such as IBM and Philips have mishandled investment into leveraging complementary product developments in the past. This is simply because they failed to consider the need for the complementor to commit to a business plan for the new product. Missing that, the new complementary product developments – dearly paid by platform owners – failed to sustainably contribute to the overall platform development (Gawer and Henderson 2007).

Smart platform development relies instead in finely crafted inducements for complementors to find their own new markets for their platform-based products/services. By doing so, the value of the platform will be substantially increased in a sustainable way, since the incentives of the complementors, the customers and the platform itself will be aligned (which is not always the case in a supply chain where one value layers could be squeezed from market power downstream in order to extract more rents).

Thus platforms must take care not to foster dependency on any "subsidy" in the firms selling platform-based complementary products / services.[4] Specifically for a platform organisation, promoting (teaching to fish) – rather than simply commissioning (giving the fish) – an innovative product/service requires a joint effort from both platform owner and complementor to create a new market. In order for this to work, solid risk-sharing arrangements must underlie this enterprise, so to provide the correct incentives to both parties.

The relationship between platform firms and complementors can be informed by how firms use hard and soft power.[5] Hard power includes threats and direct (financial)

[4] A Chinese proverb famously states: "Give a man a fish and he will eat for a day. Teach a man to fish and he will eat for the rest of his life."

[5] Yoffie and Kwak (2006) draw the notions of soft and hard power from the field of political science (Nye 2006).

incentives to manage complementors. The authors argue this can be effective in the short run but may not be sustainable in the long run – for instance by causing a backlash. On the other hand, soft power – the ability to influence by leveraging shared long-run incentives rather than immediate contractual terms – can hold profound strategic implications where a platform firm and its complementors negotiate investments aimed at market creation. The possibility for firms to add value by providing complementary services is what defines the potential a platform in enabling innovation beyond mere match-making across market sides.[6]

5.3 Governance and Risk Sharing Profitability

Because of the relationship-specific nature of the investment of a complementor in a platform, the latter is in a position of power, when the engagement between actors pans out through time. This is similar to the hold-up problem analysed in the economics literature (Joskow 1987). In a scenario where a set of complementors can only access one platform (i.e. because of their nature, location, industry), the latter could extract all value derived from the relationship, once this is established. Platform governance can therefore provide the means to address this issue by the adoption of common rules. An example of a platform rule could be that the platform extracts only the positive externalities deriving from its market-matching activities, while redistributing all remaining profits to the complementary agents – a cooperative arrangement.

The concept of governance does not only concern the exercise of market power but also involves political considerations. This is especially poignant where a platform organisation is not exclusively a commercial entity but instead holds characteristics of a governmental or non-governmental organisation (e.g. reflecting a public sector stakeholder or a cooperative charter, or an inter-firm alliance). By political we refer then to a shared understanding within the community of platform stakeholders – and of the ecosystem of firms transacting with the platform – of the division of power and responsibilities on the strategy and conduct of the platform itself.

As the literature on platform strategy has shown (Gawer and Cusumano 2002), firms who own platforms strive to establish a credible commitment towards supporting the complementor ecosystem. This is done for instance by investing in quasi-independent divisions tasked with supporting innovation in the industry, e.g. via the selective dissemination of IP, engineers/researchers time and the establishment

[6] Market creation efforts are not confined within platform strategies: many non-platform firms attempt to achieve the same goal by means of engaging and "open" marketing strategies. A separate and vast literature which highlights how key gatekeepers of networks otherwise unconnected (such as platform actors may be for different sides of the market) wield power – the so called "strength of weak ties" argument (Granovetter 1973) – addresses related issues from an actor/individual perspective.

of consultation processes at the stage of innovation design (the Intel case provides a clear illustration of this). In this context (where elements of perspectives B and C apply), the key strategic challenges for the platform are: (1) how to resolve the initial make-or-buy trade-off (make of core platform technology assets; delegate/ buy of complementary products); but also (2) how to address the incentives within its organisation so that it can both sell (its own core product) and let sell (i.e. have a commercial and legal framework enabling its complementors to develop and maintain viable businesses centred around the platform core). While the above example is set in a commercial platform context, similar strategic challenges will need to be clearly resolved also where platforms have elements of non-gov or public sector governance, addressing the key questions of who does what and how are competing incentives to appropriate value managed.

For large multinational corporations who seek a platform leadership role, this translated into the opportunity to invest resources in establishing in-house divisions whose aim is tantamount to industrial policy, only outside a governmental framework. In order to do so, these firms need to establish organisational rules and boundaries so to sustain the credibility of their commitment to supporting the whole industry, rather than undertaking exploitative divide et impera strategies. The commitment is embodied by the quasi-regulatory status of the arms-length relationship between commercially-minded units and the industrial-promotion units inside these large firms. A similar principle of division of responsibility and functional separation can ensure platform effectiveness even in contexts where the platform is not corporate.

In Italy, where very large corporations within the national industrial system are scarce, other forms of organisations could still replicate such a platform model, leveraging some of the above strategies. This model can be successful provided that the strategic rationale is focused to a key characteristics of the industrial context, i.e. the pervasive and strong economic contribution of SMEs, which can become the subject of a platform policy. We would encourage public policy stakeholders to apply platform thinking to the districts industrial milieu. A platform may be designed so to specifically engage with the characteristics of a given Italian districts and enable enhanced interaction of all district players both within and outside the district.

Given the specificity of the Italian industrial background (above all its fragmentation), platform governance is thus fundamental in ensuring that platform organisations (however established) will pursue the long-run interest of industry promotion, while maintaining a sustainable share of profit. In turn, the trust enabled by the strength of the platform commitment will provide the basis for widespread platform adoption by firms of any size.

5.4 *Innovation in Platforms: The Challenge of Market Creation*

One of the major obstacles that Italian SMEs face as they approach renewing their business models along the lines of the networked enterprise is that a relevant component of their competitive advantage derives from strong links with the firm's markets. The organisational, business and human capital changes needed in order to embrace ICT – so to transform the firm into a part of a networked system – can be seen as orthogonal to the tight market link which SMEs often perceive as the key source of their competitive advantage (e.g. supplier relationship to a major buyer, niche position, etc.). Instead, we argue that the benefits that SMEs can gain from investing in network technologies (and the organisational factors which are its complements) can be enhanced by the market creation role of platforms.

The economic and strategy literature on market creation frames this issue as a special case of technology adoption/diffusion models or in terms of new product development/marketing issues (Ghemawat 1991; Gupta et al. 1999; Parker and Van Alstyne 2005). The key question tackled is the timing of adoption of new technology, modelled over a continuum of options. The seminal paper of Fudenberg and Tirole (1985) shows how the adoption game can lead to different Pareto-efficient equilibria, which can coexist in a duopoly scenario (although not in a general oligopoly framework). In particular this could be either a delayed joint adoption or a pre-emptive outcome, where one firm introduces the new technology as first-mover. Joint adoption will result only when the first-mover imposes a large enough negative externality on the laggard. This follows from the risky nature of investment, as each firm trades-off benefits and risks of its unilateral, uncoordinated action. As Ruiz-Aliseda and Zemsky (2006) point out, this cannot be the case where adoption concerns new markets: pre-emption will then be the only optimal strategy. Firms who do not embrace the innovation first can be the clear-cut losers and forego entry to new markets. In these models the first-moving firm will be that which benefits from cost/technology advantages and risk.

We reckon that the above models, while of relevance in the case of duopolies, may not be fully predictive of incentives in those platforms which tap into existing industrial milieus, such as districts. While first movers always have some advantages, they also bear indirect costs; while the large number of followers stands to reap advantages linked to the greater maturity of the products and platforms and the greater externalities captured by a larger platform.

In general, the technology adoption perspective only provides one side of the story. Firms do not only enter markets because of access to a new technology; instead, market creation can be a distinct strategic effort, where new technologies can play an enabling, though not sufficient role. This is because the search for markets is a fundamentally uncertain process – and a pivotal one for Italian SMEs – which cannot be modelled simply as a choice to be made in a decision continuum. In fact, SMEs struggle to find the market which is the optimal match to their core competences; even where they interface with a suitable market, Italian SMEs have proven unable to fully exploit the entire set of opportunities offered by those markets. Specifically, we stress that SMEs platforms can improve the competitive position of SMEs by realigning the incentives of the large majority of services firms which could benefit from providing platform services; the resulting facilitated interaction between professional services businesses and manufacturing SMEs would in turn benefit the latter. As a consequence, an SME-oriented platform could empower SMEs – in conjunction with service providers – to create new markets for their products.

5.5 Platform Catalysts

Behind the rise of every platform there is a huge entrepreneurial effort and platform strategies play a key role in enabling success. Starting-up a platform requires a keen focus for the cross-market dynamics which a platform owner must manage. Overcoming the chicken-and-egg problem (Caillaud and Jullien 2003) which embodies the essence of market-matching, is the aim of any catalyst strategy to build platforms. Evans (2009) describes the strategies which entrepreneurs can choose in order to cajole platform support from actors belonging to different markets.

Price incentives and governance transparency can play a strong role in inducing actors to join the platform – one side of the market at a time. At the same time, platform catalysts can focus on attracting key ("marquee") customers. These are defined as customers who provide relatively larger positive externalities to users from other markets, where these join the platform. Therefore, a first strategy by which an SME-oriented platform can promote adoption is by attracting an "important" player.

The surveys on the diffusion of ICT amongst SMEs show a systematic gap between a strong SMEs online B2B practice (on purchases) and a significant reluctance to use online channels for their sales (cf. Report 3.1 on ICT adoption within Italian firms). This is explained by SMEs themselves as a result of several factors:

- No trusted third parties can provide an alternative retail channel and the SMEs' customers prefer direct face-to-face contact;
- The trust between buyer and seller is important to obtain the demanded product and ensure agreement on the exact price;
- The product is not standardised enough and must be customised on a case-by-case basis.

Moreover, where a transaction is performed online, it takes place on the website of one of the two parties, rather than on a dedicated portal. In practice, what any e-commerce arrangement fails to convey are all the transactions related to the exchange of information aiming to define the object of the transaction itself. The search for information is developed in the initial phase of scouting for companies and products as counterparts in potential transactions. Nonetheless, in a platform environment the seller's focus moves from just the customer to the transaction; customer needs will be managed as part of the transaction management above that specific customer management.

We consider that asymmetric information and knowledge barriers pose a strong barrier for the take up of professional services by SMEs which would benefit from them. Simply put, SMEs do not hold sufficient information on technologies and managerial knowledge to be able to assess the merits of innovative services which could positively complement their business expertise and competences – thus enabling greater competitive advantage. For instance, services such as the leasing of adaptive infrastructures could enable some SMEs to transform into a more cost efficient networked enterprise, which could be more effective at capturing demand and create new markets (or market niches); nonetheless, this type of solutions are currently almost exclusively the preserve of larger enterprises in absence of a network platform.

Taking this into account, we argue that a second catalyst strategy for an SME-oriented platform is to develop a key focus on initiatives aiming to develop SMEs knowledge and confidence in the opportunities offered by professional services. This role is to some extent similar to the functions of the technology/business advocates (or "evangelists" as common parlance in the US West coast), which disseminate and promote the concept and benefits of innovative products or services on behalf of high-tech firms. The key step here is realising that even in a low-tech industry context, several SMEs face a de facto knowledge barrier as to the applications of high-tech solutions (products and services) and their potential to improve the productivity of firms in low-tech industries.

Therefore, apart from carefully crafting market incentives on both side of the markets, we envisage that successful SMEs-oriented platforms will engage in information dissemination and persuasive confidence building strategies (e.g. pilot projects) to catalyse platform adoption. This "evangelist" catalyst role can be pivotal at the start of a platform take up; when sufficient momentum has been gathered, this catalyst effort can be turned to another platform with a different remit (e.g. a different type of services) which is in the early stages of adoption. One way to fund this catalyst function would be by allocating a share of the platform transaction commissions resulting from its actions – as well as perhaps a policy subsidy devised as an incentive for the platform to fully develop and internalise the cross-market externalities (rather than leave some SMEs sub-markets underserved).

Finally, we note that a competitive advantage for an SME platform may lay specifically in harnessing trust from its subscribers and define itself as an "honest broker", which can be relied upon to clarify the interests of the parties. For this reason, a platform where the key characteristic is its matchmaking role and its transparent pricing schedules (akin to the Platform A archetype) could be successful by enabling interaction between the parties by means of specialised and standardised transactions solutions (products/services). At the same time, a platform which holds IP which the SMEs would want to build upon/incorporate as part of their product or service development (this could be the case where a Platform B perspective is applicable) – because of its incentives to "enlarge the pie", i.e. increase the diffusion of its core platform product by enhancing the range of solution, services and goods which incorporate the core product – could also be seen as a balanced mediator of the interests of parties located on different sides of the market. On the other hand, trust is a key challenge where a (large) company opens up its supply chain (Platform perspective C), which is why the devising of appropriate governance mechanisms can contribute dramatically to the success of this type of platform endeavours. For this reason, public policy intervention setting benchmarks and frames of reference for platform governance (akin to regulations of consortia) could play a beneficial and very effective role. These considerations will inform our discussion of a proposed policy approach for SMEs services platforms in the next section.

6 A Proposed Policy of Service Platforms for the Renewal SMEs

The current absence of platforms which aim to serve the interest of SMEs can be associated to the structural characteristics of both the manufacturing industries and also in the service industries which do not currently include value creation services for SMEs as a relevant stream of business in particular in Southern Italy. For instance, while the Italian banking system has introduced network services used by SMEs, the services themselves are designed so to reflect the needs of the banks themselves rather than elicit the set of needs which SMEs could satisfy with an advanced tailored e-banking platform. This follows from the joint efforts by multiple parties within the banking system to create a shared infrastructure and electronic services base, funded exclusively by the banks.

Targeted policy intervention embracing all services domains could potentially alter the status quo and bridge the existing market gap/failure. In fact, Gawer (2009) suggests that evolutionary trajectories may define which types of platform characterise an industry: under certain conditions, internal platforms can evolve into supply-chain platform, which on their turn can become business platforms. Thus, policies aiming to ameliorate the competitiveness of SMEs via the promotion of supply-chain platforms may want to target existing internal platforms (i.e. proprietary to a firm's product/service design) and devise the incentives for these platforms to be shared across the supply chain. Note that this process could be detrimental to the firm currently holding the internal platform (see the analysis of the IBM case in Gawer 2009). While internal platforms that are based on products rely on design skills and are protected by the system-maker via formal intellectual property such as patents, those based on services – the key target of a policy aimed to enhance the competitiveness of SMEs – will be dependent on "softer" ways (including the use of existing sectoral and geographic relational networks) to defend competitive advantage at the systemic level, such as trade secrecy, business contacts, etc.

We argue therefore that – by taking into account the characteristics of the diverse range of platforms or quasi-platforms which already operate across many markets – it is possible to conceive and define a set of platform specifically targeting the SME needs. Specifically, we consider that SMEs can engage in different ways with platforms. A first approach is to consider a platform which allows firms (predominantly large firms) to trade with each other, perhaps facilitating international trade. This may be the case in clusters in the automotive or ship building industries. This platform may compete with an alternative distribution channel, where a merchant will own the suppliers' stock and resale at

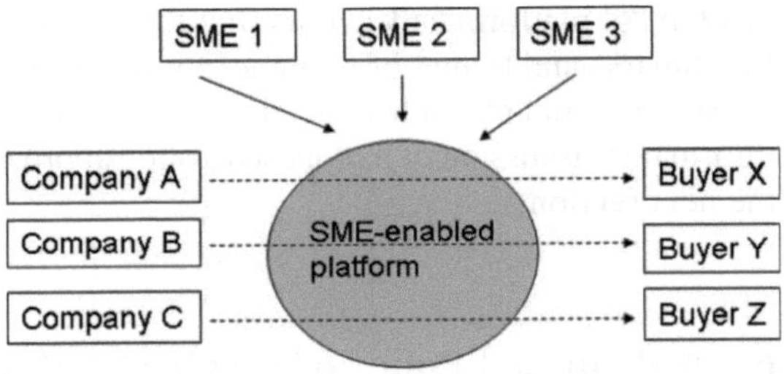

Fig. 1 The SME-enabled platform approach

a margin to the buying firms. Consider an SME that can provide value added services (which could be embodied in a service associated with the products traded by the large firms). This SME will have to overcome several legal and commercial constraints in order to be able to sell its services atop the main goods traded by the merchant and its distribution channel.

On the other hand, the same SME – if engaged with a trading platform which incorporates by design the provision of complementary services atop the main traded goods – could draw on the (large) buying firms' incentives to select as trading partner only (large) supplying firms which enable the complementary services. Large suppliers, which benefit from the extra platform trade, have an incentive to allow SMEs services to complement their goods as they may be at a competitive disadvantage against other suppliers on the platform who do so.

Finally, this platform may be competing with other trading platforms which may not enable SMEs engagement; if SMEs products/services bring net value added to the platform, then this can provide a source of competitive advantage to those platforms who tap into their potential. We will refer to this as the SME-enabled platform approach (cf. Fig. 1).

The services provided by SMEs are not directly (but only indirectly) of benefit to Company A, B or C, while they are directly of interest for the buyer. Company A (or B, C) will benefit by enlarging the pool of buyers by accessing a broader set of SMEs. This simple platform model involves in fact a more transparent interaction between the end user and the SME which provides products/services which are complementary to the key product offered by Company A (or B, C). The key challenge here is to enlarge the ultimate target market of a given SME, by allowing several (large, pivot) Companies to operate on the same platform – where previously the ultimate users of that SME's output were channelled by the supply chain of a single (large) Company. For this reason, for this type of platform model of interaction to take off, it is necessary for both SMEs and (large) Companies to accept a greater degree of immediate competition. At the same time the total degree of competition would remain unchanged, since in the present "flat" world different supply chains and platforms (including from emerging countries) are already challenging the supply chains and platform of OECD countries across a significant range of industries.

We now propose an alternative approach which we reckon to be complementary (rather than exclusive) to the above platform model. In this case, we consider a

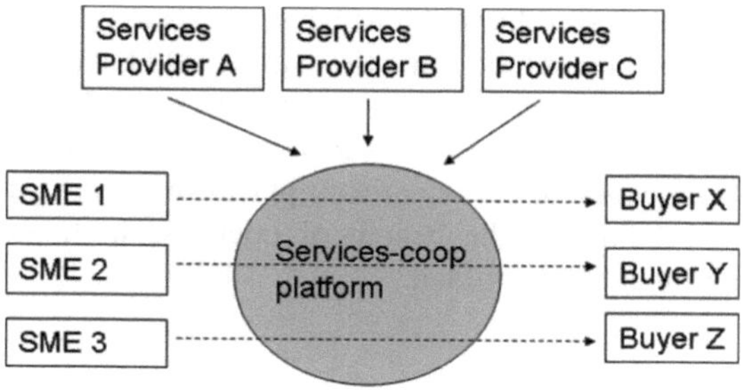

Fig. 2 The services-coop platform approach

platform where SMEs supply the main goods, seeking trade with buyers (e.g. international buyers). Here, the platform not only intermediates between the SMEs and their (new) target markets, but also enables suppliers of professional services (marketing, communication, finance, ICT consultants) to come in and engage with SMEs. Thus, by aggregating SMEs, the platform provides a mutually beneficial opportunity for professional services firm to tap into underserved SME markets, simultaneously allowing SMEs to invest in these value-adding professional services to improve the service-value of their goods. By similarity with the established wine cooperatives (commonplace in Italy, amongst other countries), we define this as the services-coop platform approach (cf. Fig. 2).

In this case, the key feature of the platform is the aggregation of similar SMEs (by co-opetition, i.e. at a pre-competitive stage), which allows a given SME to mimic the economies of scale and scope of larger players in order to be able to access a set of professional services from which it would otherwise be de facto excluded. For instance, Italian SMEs at present generally have access to the professional services of accountants and retail commercial bank lending though not services advising/procuring ICT transformation or marketing. A platform bridging the gap between (unexpressed) demand and (uninterested) supply of these latter categories of professional services, could enable Italian SMEs sellers, by leveraging the input of Italian or multi-national professional services companies, to interact with buyers in new markets such as for instance China.

It ought to be noted that one of key features of the services-coop platform approach is its market creation role, which as we have discussed above is an important opportunity associated to some platforms. In this case the first market creation lies in the opportunity of enabling professional services firms to complement and add value to SMEs products (e.g. by enhancing their marketing or communication potential). By matching these (currently unmatched) set of firms, the SMEs may then also be able to develop new markets for their products.

Moreover, a services-coop platform can ensure that a leading professional services supplier commits to engage with SMEs via the platform. This could have a first impact on the number and type of SMEs joining the platform; at a second stage, the presence of the leading service company could induce some of its competitors to join too to serve the SME market. This confirms that for an SME services adoption initiative to achieve maximum impact, exclusivity policies may be counterproductive – which suggests that an open platform approach may be the best policy. At the same time, this could have implications on governance: an open platform may be more likely to succeed where multiple stakeholders share risks, oversight responsibilities and rewards from the platform. On the contrary, a platform captured by a dominant player (even if a public sector organisation) may end up trading-off the benefit of openness in favour of a particular interest of the dominant stakeholder.

6.1 The Features of a Service Platforms Policy for SMEs

Building upon the analysis presented throughout this work, we now proceed to elucidate a proposal to leverage the platform industrial organisation in order to achieve a policy objective: promote the adoption of advanced ICT solutions amongst SMEs. Specifically, we aim to fine-tune this proposed policy approach so to target SMEs which achieved competitive specialisation in legacy low-tech industries – such as the case of Italian SMEs and those in many low-tech clusters, districts or filières.

We have highlighted how SMEs can benefit from a platform organisation of industry in different ways. First, SMEs can tap into the transactions between **large firms** occurring within an SME-enabled platform. Here SMEs are the players providing the complementary, value-adding services to the suppliers of the principal product.

Alternatively, SMEs – which engage in direct trade with their buyers – can benefit from services platforms. These, defined above as services-coop platforms (cf. Fig. 2), provide a medium for SMEs to engage with **professional services suppliers**, which can greatly enhance the service value component of the SMEs offering (i.e. manufacture).

For instance, a set of SMEs belonging to the same industry or in close geographic proximity (within a district for instance) may leverage the platform to share access to an adaptive infrastructure, maintained by the ICT services provider. By doing so, each of the SMEs could innovate around its existing product and competences, without having to suddenly overcome a huge ICT knowledge and management barrier. In turn the service provider would likely benefit from the demand aggregation function performed by the platform and fine-tune an array of solutions for SMEs.

We scope our proposed policy development around this latter type of platform. What we would like to advocate is for policy-makers to consider promoting the advent of services-centric platforms targeted at SMEs. In order to do so, a policy could be envisaged so to provide incentive for professional services firms (services beyond banking) to coordinate and engage with SMEs.

As discussed above, the key to a platform's success in gathering sufficient momentum to become a sustainable business is how effective it is at catalysing the latent interest amongst parties at all sides of the market (SMEs, professional services suppliers, to some extent also buyers of the final product which may be drawn by a platform certification scheme or the higher profile associated to the input of the professional services firms). Therefore policy intervention, rather than attempting to devise a catch-all subsidy (unlikely to be as effective as expected and perhaps legally incompatible with EU laws, as considered below) could focus on support for a set of agents to mobilise SMEs interest in the professional services supplied on the platform and garner trust in the overall remit and functioning of the platform – thus providing the essential catalyst role.

Another pivotal objective of a public policy aimed at promoting platform based services (including ICT) for SMEs is the creation of a set of appropriate laws and regulations which are sufficiently clear and flexible so to elicit effective governance regimes within platforms. In fact, both during and after the catalyst stage, trust in the platform (especially by the smaller and weaker players) will depend on how transparent the governance is, i.e. the division of responsibilities, risks and rewards within the platform. Therefore a platform-specific legal framework could be an essential policy input for a fast and sustained establishment of services centric platforms for the benefit of SMEs.

6.2 *Which Policy and for Which SME?*

Public policy needs not disrupt industry equilibria but may instead align itself with industrial dynamics trends. In the general domain of industrial policy promotion, the literature on state subsidies shows how SMEs subsidies could have had unintended negative effects due to adverse selection and moral hazard (Colombo and Grilli 2006, 2007; Santarelli and Vivarelli 2002): firms who rely on state aid and subsidies without more than a passing interest in viable commercialisation do not contribute to sustainable innovation.

Carefully crafted public policies could thus ensure that the positive economic externalities from this process can be reaped in part by the internal platform firm, so to rebalance its otherwise conflicting proprietary incentives. Public intervention can enable healthy platform governance if it provides the right rules to prevent anti-competitive behaviour within platform. This could be the case if one of the transaction parties on the platform can act as a monopolist and appropriate in its entirety the value generated by other parties (such as SMEs).

We would like to stress that a platform-specific legal framework would likely involve provisions that ensure that under all circumstances:

- SMEs serving different products on different markets can participate to the same platform; and
- A stakeholder mechanism ensures SME representation on the platform management board; and
- A defined share of the platform-generated revenues filters back to the SMEs.

7 Concluding Remarks

This work builds upon a review of the concept of platform within the economics and technology and management strategy literature in order to identify a positive policy approach to ameliorate shortcomings in ICT investment by SMEs. We have identified across this literature three definitions of platform, each of which stresses the importance of their role in:

- Managing transactions across different market sides (linked by cross-market externalities and platform prices): **Platform A – Economics perspective**.
- Enabling the creation of new services and markets by promoting the adoption of a core platform technology and coordinating technological development and the necessary standards: **Perspective B – the platform as a standard**.
- Opening the supply chain of a set of key (e.g. pivot) firms, enabling a greater degree of independent interaction between these firms' suppliers (e.g. SMEs) and (a broader set of) end users of the products: **Perspective C – the platform as an open supply chain**.

We have highlighted how SMEs can benefit from a platform organisation of industry in different ways. First, SMEs can tap into the transactions between large firms occurring within an SME-enabled platform. In this case, SMEs provide value added products and services which are complementary to the main product exchanged via the platform. This is similar to the function already performed by many SMEs under currently established industrial structures, where SMEs deal exclusively via a pivot firm, the large supplier of the main product. Note that the latter can appropriate a share of the value provided by the SME if supplied under a vertical rather than platform industrial structure. At the same time a pivot firm may be missing out on the benefits of having its product complemented by products/services provided by SMEs other than those in close proximity to the pivot firm's cluster or filière. Therefore both large and small firms can potentially benefit from a platform approach, given its win-win nature.

As a next step we have focused on services-coop platforms and analysed how this approach to industrial organisation can enable SMEs to benefit from access to professional services inputs in order to enhanced their products and capabilities. By way of example, a services-coop platform could enable specialists in the installation and strategic management of ICT (such as adaptive infrastructures) to engage with multiple SMEs, by aggregating their demand.

Moreover, we stress that a services-coop platform would require efforts to be directed towards catalysing interest, support and trust in the platform and in the technology and services provided. This is an area where we urge strong and timely public policy support, which we believe could have considerable impact – even with limited public expenditure – by carefully though effectively realigning (or "nudging", as in the latest policy parlance) the incentives of several types of firms (SMEs in primis, but also professional services firms and larger suppliers and buyers) so to ultimately promote a self-supporting ecosystem of platforms for

SMEs spanning (and bridging) the present day ecosystem of industrial districts which characterises economies such as Italy.

In this paper we have described a growth model for the Italian industry and services SMEs focused on technology and inter-firm collaboration based on a platform technology to expand in foreign markets. The extant policy advice induces SMEs to defend their competitive position by reducing labour costs in an ineffective and unnatural attempt to match the features of emerging countries' firms which are now taking a large share of the export markets traditionally served by the Italian SMEs. The model proposed in this paper, instead, aims to persuade our SMEs to conquer new markets and market niches through platform collaboration which involves the participation of suppliers, customers and complementary services providers, which – by interacting on the platform – can increase the value of the output without (some of) the zero-sum game limits of traditional supply chains.

References

Acconcia A et al (2009) Adozione e diffusione delle Ict nelle imprese manifatturiere italiane. In: Rondi L, Silva F (eds) Produttività e cambiamento nell'industria italiana. il Mulino, Bologna

Antonelli C (1997a) A regulatory regime for innovation in the communication industries. Telecommun Policy 21(1):35–45

Antonelli C (1997b) New information technology and the knowledge-based economy. The Italian evidence. Rev Ind Organ 12(4):593–607

Armstrong M (2006) Competition in two-sided markets. Rand J Econ 37(3):668–691

Bresnahan TF, Trajtenberg M (1995) General purpose technologies as 'engines of growth'? J Econometrics 65:83–108

Caillaud B, Jullien B (2003) Chicken & egg: competition among intermediation service providers. Rand J Econ 34(2):309–328

Chesbrough H (2011) In: Di Minin A (ed) Open services innovation competere in una nuova era, Sxi Springer per l'Innovazione. Springer, Milano

Colombo M, Grilli L (2006) Supporting high-tech start-ups: lessons from Italian technology policy. Int Entrepreneurship Manage J 2(2):189–209

Colombo MG, Grilli L (2007) Technology policy for the knowledge economy: public support to young ICT service firms. Telecommun Policy 31(10/11):573–591

Evans DS (2009) How catalysts ignite: the economics of platform-based start-ups. In: Gawer A (ed) Platforms, markets and innovation. Edward Elgar, Cheltenham/Northampton, Chap. 5

Ferdows K, Lewis M, Machuca J (2004) Rapid-fire fulfillment. Harvard Bus Rev 82(11):104–110

Friedman TL (2006) The world is flat. The globalized world in the twenty-first century. Penguin books, London

Fudenberg D, Tirole J (1985) Preemption and rent equalization in the adoption of new technology. Rev Econ Stud 52(170):383

Gawer A (2009) Platform dynamics and strategies: from products to services. In: Gawer A (ed) Platforms, markets and innovation. Edward Elgar, Cheltenham/Northampton

Gawer A, Cusumano MA (2002) Platform leadership: how Intel, Microsoft, and Cisco drive industry innovation. Harvard Business School Press Books, Cambridge, MA

Gawer A, Henderson R (2007) Platform owner entry and innovation in complementary markets: evidence from Intel. J Econ Manage Strategy 16(1):1–34

Ghemawat P (1991) Market incumbency and technological inertia. Market Sci 10(2):161–171

Giannakouris, M. Smihly (2010) ICT usage in enterprises. In: Data in focus, EUROSTAT, Luxembourg

Granovetter M (1973) The strength of weak ties. Am J Soc 78(6):1360
Gupta S, Jain DC, Sawhney MS (1999) Modeling the evolution of markets with indirect network externalities: an application to digital television. Market Sci 18(3):396–416
Hagiu A (2006) Pricing and commitment by two-sided platforms. Rand J Econ 37(3):720–737
Hagiu A (2007) Merchant or two-sided platform? Rev Netw Econ 6(2):115–133
ISTAT (2011) Le tecnologie dell'informazione e della comunicazione nelle imprese, statistiche in breve. ISTAT, Roma
Joskow P (1987) Contract duration and relationship-specific investments: empirical evidence from coal markets. Am Econ Rev 77(1):168–185
Nye J (2006) Soft power: the means to success in world politics. Perseus Books Group, New York
Parker GG, Van Alstyne MW (2005) Two-sided network effects: a theory of information product design. Manage Sci 51(10):1494–1504
Rey GM (2012) Once upon a time ...: a reflection on the relationships between industry and services. Economia Internazionale, 1, vol. 65 (pp.109–147)
Rey GM, Varaldo R (2011) Crescita economica e rinnovamento imprenditoriale. Econ Ital 3:741–781
Rochet JC, Tirole J (2003) Platform competition in two-sided markets. J Eur Econ Assoc 1(4):990–1029
Rochet JC, Tirole J (2006) Two-sided markets: a progress report. Rand J Econ 37(3):645–667
Ruiz-Aliseda F, Zemsky PB (2006) Adoption is not development: first mover advantages in the diffusion of new technology. SSRN, sSRN eLibrary
Sallusti F (2010) Relazioni fra imprese e reti: analisi e studi di caso. L'Industria 1:83–113
Santarelli E, Vivarelli M (2002) Is subsidizing entry an optimal policy? Ind Corp Change 11(1):39–52
Vargo SL, Lusch RF (2004) Evolving to a new dominant logic for marketing. J Market 68(1):1–17
Weyl EG (2010) A price theory of multi-sided platforms. Am Econ Rev, Volume 100, Number 4, pp. 1642-1672(31)
Yoffie DB, Kwak M (2006) With friends like these. Harvard Bus Rev 84(9):p88–p98

Starting Smaller; Staying Smaller: America's Slow Leak in Job Creation

E.J. Reedy and Robert J. Strom

1 Introduction

The United States, like many developed economies, is in the midst of a jobs crisis. While the recession was officially declared over in June 2009, the U.S. unemployment rate at this writing in July 2011 is still above 9% of the active labor force. If the under-employed are counted in this total, the employment deficit is closer to 16%.[1]

Consider these disturbing facts:

- In 2010, 11% of people who found a job after being unemployed had been out of work for more than a year (almost four times the level in 2007 when 3% were unemployed for 12 months or longer).
- In 2010, 34% of people who found work after being unemployed had been jobless for less than 5 weeks (down from 49% in 2007).[2]
- Currently, the percent of the population working has fallen almost 5% points from 63.1% to 58.4%.[3]

Recent Census Bureau research has pointed to one factor that is contributing to this slowdown in job creation – shrinking job creation in startups. As shown in the Fig. 1 below, startups created an average of 3.5% of total U.S. jobs annually in the 1980s, but in the 2000s contributed only 2.6% of total U.S. jobs. While diminished in number, these jobs were still the difference between positive and negative overall net job growth in the United States.

[1] See Izzo (2011).

[2] See U.S. Bureau of Labor Statistics (2011).

[3] See Samuelson (2011).

E.J. Reedy (✉) • R.J. Strom
The Ewing Marian Kauffman Foundation, 4801 Rockhill Road, Kansas, MO 64110, USA
e-mail: ereedy@consultant.kauffman.org; rstrom@kauffman.org

G. Calcagnini and I. Favaretto (eds.), *Small Businesses in the Aftermath of the Crisis*, Contributions to Economics, DOI 10.1007/978-3-7908-2852-8_4,

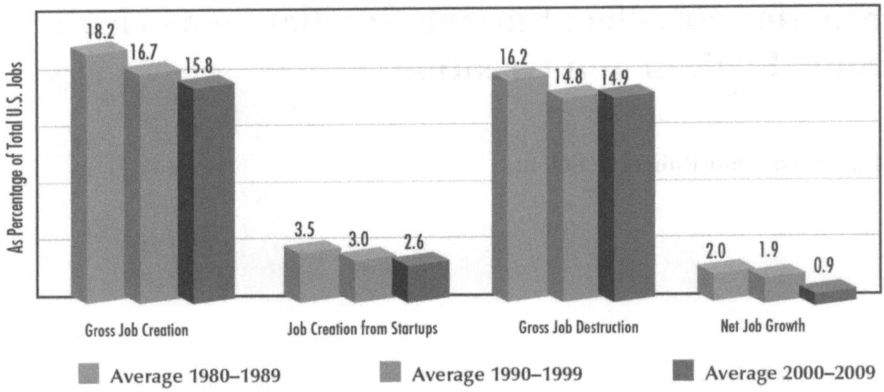

Fig. 1 Trends in gross flows and net job creation (Source: Business Dynamics Statistics Briefing: Historically Large Decline in Job Creation from Startup and Existing firms in the 2008–2009 Recession, Kauffman Foundation, March 2011, http://www.ces.census.gov/docs/bds/plugin-BDS%20March%202011%20single_0322_FINAL.pdf)

Media and academic commentators who bemoan America's unusually slow rate of job creation after the 2007–2009 recession are missing what we believe is a longer-term trend that began earlier in the decade and might best be called a slow jobs "leak." In the pages that follow, we draw upon newly available data to track businesses over time and dig deeper into the health of U.S. startups. We examine young companies' size at birth, jobs created, and survival patterns to draw inferences about the health of emerging companies in the United States. The patterns we find among young businesses show that recent U.S. startups are performing much worse than prior cohorts in terms of job creation.

Conventional wisdom about job growth tends to focus solely on the jobs that are being created at existing (typically big) companies. But as a wealth of recent research has shown,[4] new businesses are vital contributors to a healthy jobs market. Indeed, we know that until the Great Recession, new firms in the United States generated on average about three million new jobs every year. While these firms typically follow a quick up-or-out pattern of success or failure, our analysis highlights for further scrutiny some additional and, we believe, significant facts about the jobs actually created by new businesses.

2 Note on Data Sources Used

In this paper, we primarily draw on data that are a part of three government time series. One such series comes from the U.S. Census Bureau's Business Dynamics Statistics program,[5] which allows for the tracking of new, independent business

[4] See, for example, Haltiwanger et al. (2010).

[5] Data by firm age pulled from http://www.ces.census.gov/index.php/bds/bds_database_list.

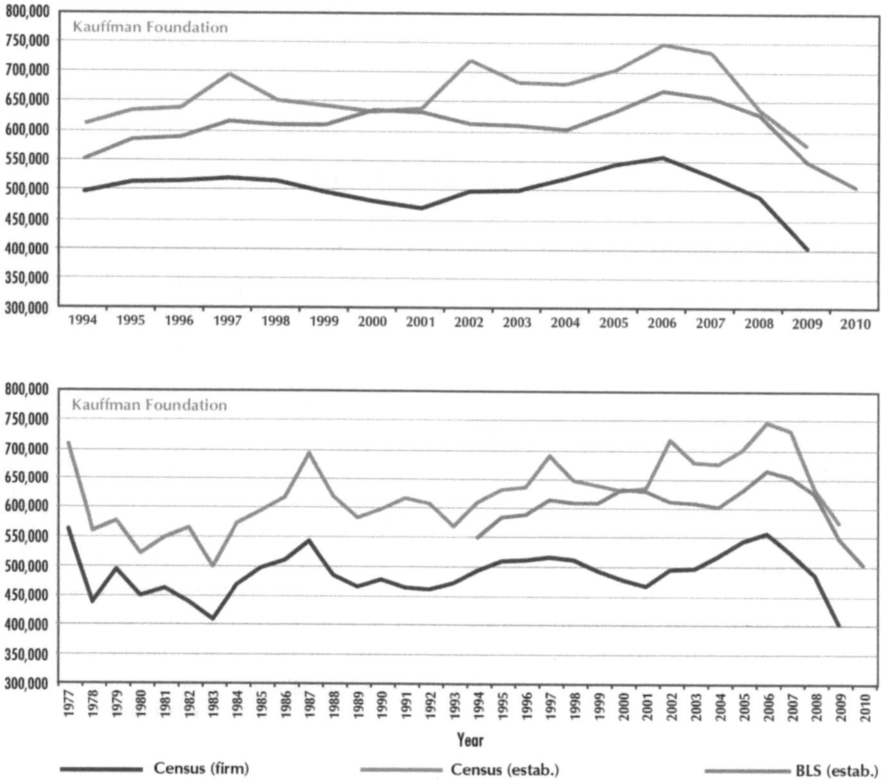

Fig. 2 U.S. business births (firms and establishments) (Source: Author Calculations from Census Bureau's Business Dynamics Series (published series and unpublished establishment portions of the series) and Bureau of Labor Statistics' Business Employment Dynamics Program)

starts or firms. But because of differences observed between these data and other more timely data, we also present establishment data from the U.S. Bureau of Labor Statistics' (BLS) Business Employment Dynamics program[6] and unpublished establishment data from the U.S. Census Bureau's Business Dynamics series.[7]

[6] Data by firm age pulled from http://www.bls.gov/bdm/us_age_naics_00_table7.txt.

[7] While this data on establishment age dynamics had cleared the Census Bureau's disclosure process, it was not published by the Census Bureau at the time of this research and was provided to assist in the investigation of some of the differing trends seen in the series.

Although the series have some surface contradictions, until it is clear which series is the most informative, we believe it best to present data based on all three time series and note the significant differences where they exist.

3 Trends in Business Starts and Survival

Recent data on business dynamics from BLS and the Census Bureau have allowed researchers to closely track business startups. Especially important are "employer-based" firms because these businesses not only generate jobs for other workers in addition to their entrepreneurs, but also are more likely to scale to become a continuing source of job creation.

As shown in Fig. 3, prior to the current economic crisis, the number of new employer businesses – those that provide work for individuals other than the founder[8] – mostly held steady or meandered upward over the several prior decades.[9] But, as shown in Fig. 2, even before the overall economy started its most recent downturn, the number of new employer business births in the United States had peaked.

In fact, since 2006, the annual number of new employer businesses tracked in government statistics has plummeted 27% for new independent firms and approximately 23% when considering a broader measure of new "establishments" (new independent businesses *and* new expansions of existing businesses).[10]

Another important, but often overlooked, measure is the rate of business survival for all new firms as depicted in Fig. 3.

Somewhat remarkably, the overall trend in this statistic was relatively stable prior to the Great Recession, with roughly 45–50% of new businesses surviving 5 years. Recent business births, however, are showing starkly lower survival rates. The survival data, whether measured at 2 years (61% for firms born in 2007, compared to about 65% for 2-year-old firms in earlier periods) or 5 years (now under 45% for firms born in 2004), indicate that recent cohorts of new businesses have been very hard hit by the 2007–2009 downturn.

[8] A further discussion of the distinctions of employer and nonemployer firms and their relative sizes and impacts on the economy is available here: http://www.census.gov/econ/smallbus.html.

[9] The Kauffman Index of Entrepreneurial Activity and the Small Business Administration's annual tallies are but two examples. These two measures are discussed indepth in other recent papers that are a part of Kauffman's Firm Formation and Economic Growth Research Series (http://www.kauffman.org/research-and-policy/firm-formation-and-economic-growth-research-series.aspx).

[10] This trend differs significantly from Kauffman's own Kauffman Index of Entrepreneurial Activity (KIEA). The KIEA includes all types of business starts, both sole proprietorships and those firms engaging employees. The KIEA data suggest that the increase in the overall number of new firm starts in recent years has come from a rise in the former, not the latter.

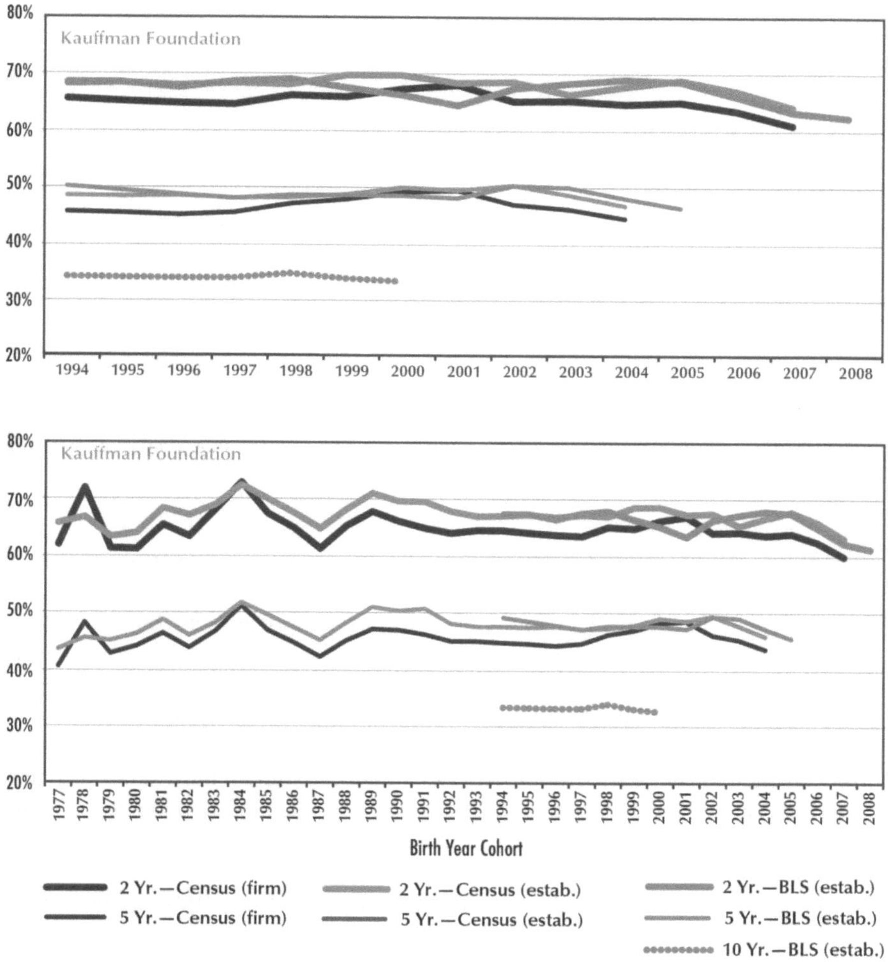

Fig. 3 Average survival rates of U.S. new business cohorts (firms and establishments) (Source: Author Calculations from Census Bureau's Business Dynamics Series (published series and unpublished establishment portions of the series) and Bureau of Labor Statistics' Business Employment Dynamics Program)

4 Employment Trends

So far, we have looked only at the numbers of new businesses and their overall survival rates. What about the numbers of *jobs* that new businesses have created, and what are the trends in these data?

Because of unknown differences in the data sources from Census and BLS – a mystery we urge other researchers to tackle – employment within business

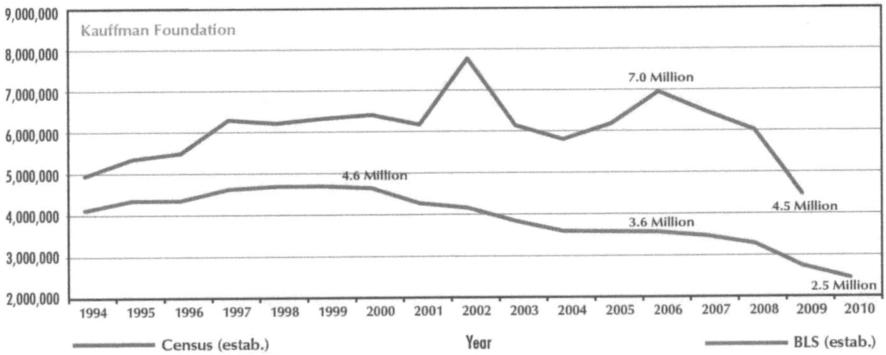

Fig. 4 Total aggregate employment of U.S. new business cohorts (establishments) (Source: Author Calculations from Census Bureau's Business Dynamics Series (published series and unpublished establishment portions of the series) and Bureau of Labor Statistics' Business Employment Dynamics Program)

cohorts is somewhat cloudy. The Census Bureau's data on employment appear significantly more positive than the BLS data. But rather than attempt to sort out the differences between the two series here, we simply present both, and urge readers to recognize that, despite the differences, the overall story in both series is disturbing.

To begin, we look at the aggregate numbers of jobs reported by all new businesses and new *establishments* (new plants or offices opened by existing firms) in their birth year. This aggregate measure includes more than just new firms, but is nonetheless a useful indicator of the dynamism of the economy (Fig. 4).

Figure 4 indicates that employment at new establishments, as measured in the BLS series, has moved steadily downward since about 2000. The data show an average peak of about 4.7 million new jobs [annually] from 1997 to 2000, compared with fewer than 2.5 million jobs at new establishments in 2010. Comparable Census figures show a much later – but steeper – decline in aggregate employment at new establishments, peaking in 2006 at just under seven million jobs and then plummeting to less than 4.5 million by 2009. In the end, and despite the differences in timing, the two series show very similar annual declines of 2.2 million and 2.5 million new jobs, respectively, generated by new business establishments compared to recent peaks.[11]

Census also provides data on jobs created by new independent firms only (a subset of the establishment measure), which are displayed in Fig. 5. These data also show a significant decline in job creation. According to the Census data, new

[11] It is expected because of industry differences that these two series should have some variations in levels; the differing nature of when they begin to measure employment declines was not expected.

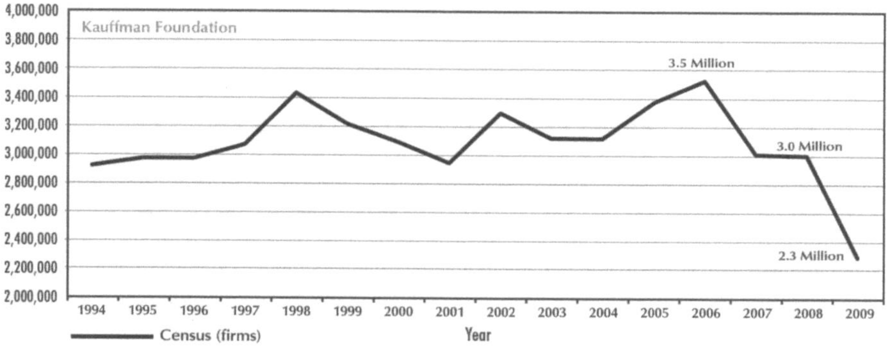

Fig. 5 Total aggregate employment of U.S. new business cohorts (Firms) (Source: Author Calculations from Census Bureau's Business Dynamics Series (published series and unpublished establishment portions of the series) and Bureau of Labor Statistics' Business Employment Dynamics Program)

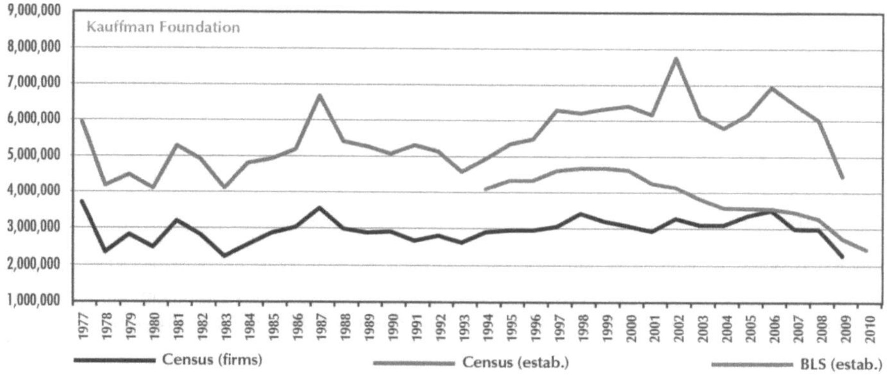

Fig. 6 Total aggregate employment of U.S. new business cohorts (establishments and firms) (Source: Author Calculations from Census Bureau's Business Dynamics Series (published series and unpublished establishment portions of the series) and Bureau of Labor Statistics' Business Employment Dynamics Program)

firms generated steady job contributions of more than three million jobs per year through 2008, but that number has since fallen substantially to 2.3 million in 2009, a decline of 700,000 jobs.

The big trend is unambiguous across all of the data series: *New business births have been contributing fewer jobs to the economy in recent years*. Aggregate employment is suffering not simply because of falling numbers of new business starts, but also because average employment at each new firm is shrinking (Fig. 6).

In Fig. 7, BLS data show that new establishments opened their doors with about 7.5 jobs on average for much of the 1990s, a figure that has since declined to 4.9 jobs per new establishment. The Census data on establishments show a similar

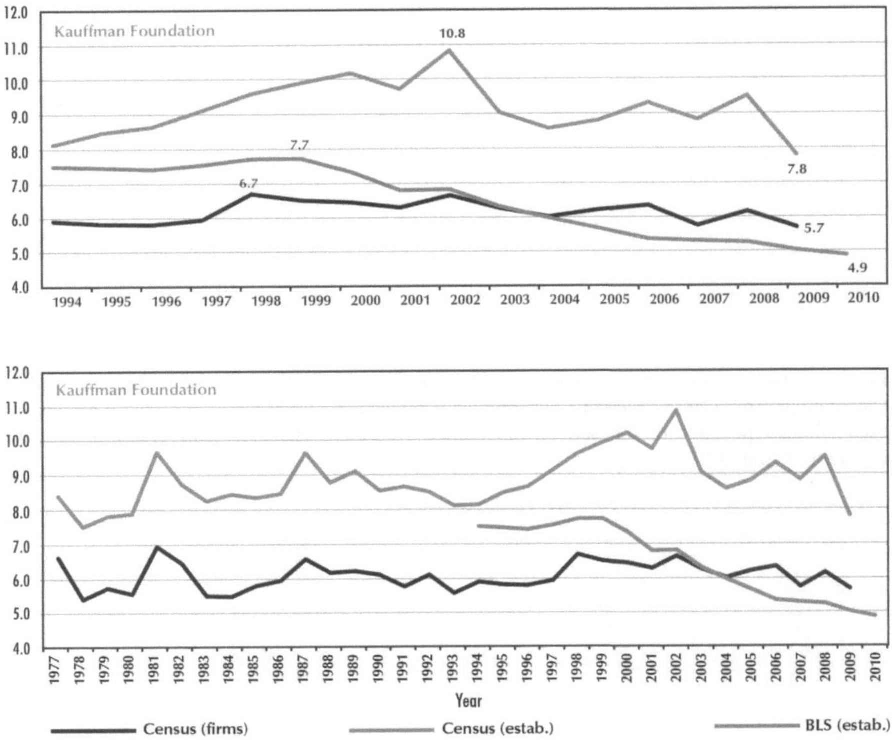

Fig. 7 Average startup size (establishments and firms) (Source: Author Calculations from Census Bureau's Business Dynamics Series (published series and unpublished establishment portions of the series) and Bureau of Labor Statistics' Business Employment Dynamics Program)

pattern over this same period, but with different levels: Establishment startup size actually increased steadily from about eight employees per business to a peak of 10.8 in 2002, before falling to a steady range of about nine employees in the mid-2000s and then declining again in 2009 to fewer than eight employees per establishment. On the other hand, the firm-level statistics at Census show only a modest decline from a level of about six employees per firm over most of the last two decades.

The aggregate and per-business numbers paint two similar narratives of why new businesses have been contributing fewer jobs. This falling contribution reflects both a drop in the number of new establishments *and* shrinking employment levels at these establishments. In both data series, the most recent year of data shows a cohort of new business that was smaller in number and in jobs created than in any cohort since 1994 and, in most cases, than *any* previously measured cohort in data dating back to 1977.

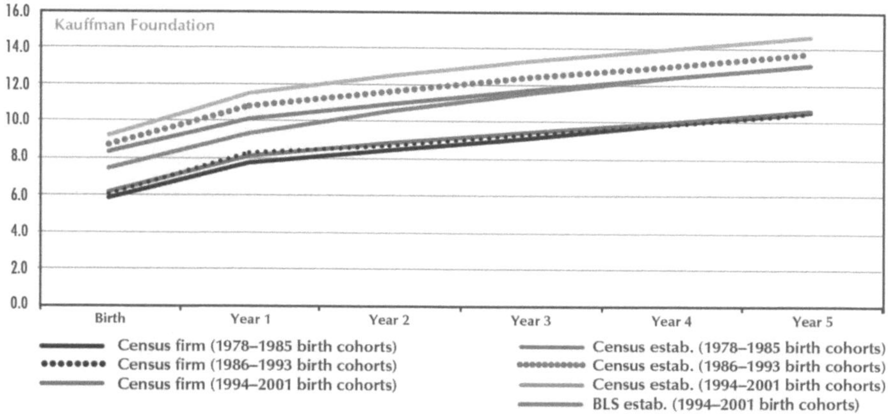

Fig. 8 Average employment of surviving businesses as they are born and age to year 5 (Source: Author Calculations from Census Bureau's Business Dynamics Series (published series and unpublished establishment portions of the series) and Bureau of Labor Statistics' Business Employment Dynamics Program)

5 Explaining the Employment Trends at New Businesses

One hypothesis that could be advanced about recent new business starts is that, while they may be starting smaller (or leaner), as a group they will grow more rapidly over time. Indeed, there is ongoing debate over whether this hypothesis explains the different patterns in entrepreneurship between Europe and the United States.[12]

Fortunately, new data from Census and BLS permit some testing of this hypothesis by tracking employment and survival of each cohort of new businesses over time. Business cohorts – all the businesses born in a particular year (and a particular circumstance) – are just like birth cohorts or generations; they tend to have some shared characteristics or imprinting. And, indeed, just like children that age, businesses that survive their early years tend to grow, on average, such that the following picture of average employment at a business over time appears (Fig. 8).

In looking at the average employment of surviving firms or establishments through these growth charts, it should be noted that establishments and firms drop out of the cohort over time as they are acquired, close, or otherwise cease to exist. What Fig. 8 shows, in all the data series and for both firms and establishments, is that those businesses that remain in operation tend to get bigger, not smaller, as measured by the number of people they employ.

[12] See Bravo Biosca (2010).

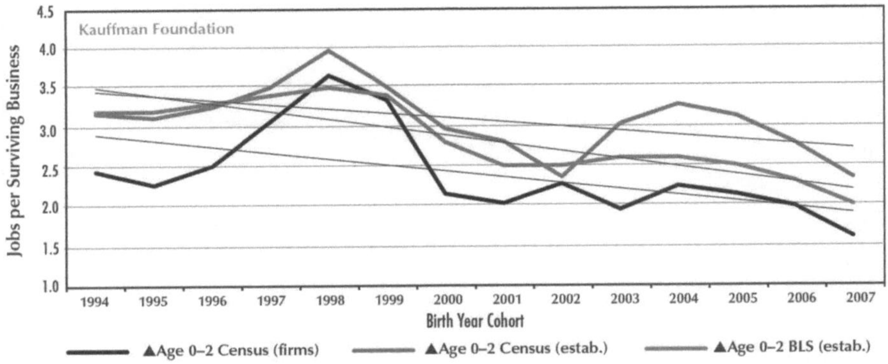

Fig. 9 Change in size of surviving business from birth to age 2 (establishments and firms) (Source: Author Calculations from Census Bureau's Business Dynamics Series (published series and unpublished establishment portions of the series) and Bureau of Labor Statistics' Business Employment Dynamics Program)

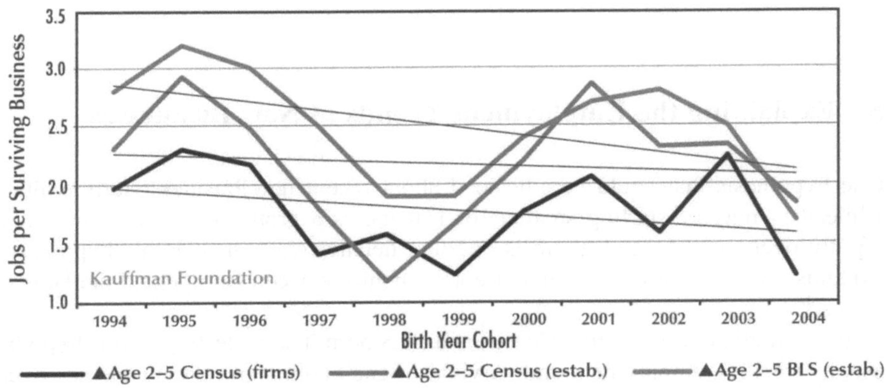

Fig. 10 Change in size of surviving business from age two to age 5 (establishments and firms) (Source: Author Calculations from Census Bureau's Business Dynamics Series (published series and unpublished establishment portions of the series) and Bureau of Labor Statistics' Business Employment Dynamics Program)

We have broken the available time series into three distinct periods to show the similarities and also some differences in the trends over time. These growth charts allow us to expand our understanding of how young firms contribute to jobs as they age. Just as we presented some differences in how the BLS and Census series depict recent trends in employment, we see differences in the growth charts. But, more important, in our view, are the similarities the data show. Specifically, we want to focus on the downward trends seen in all the series in the average rate of growth for new businesses since 1994.

Figures 9 and 10 above present some mixed news. They show that, while new businesses that survive continue to be significant contributors to employment as they age and grow, the rate (or slope of the trend line) of their employment

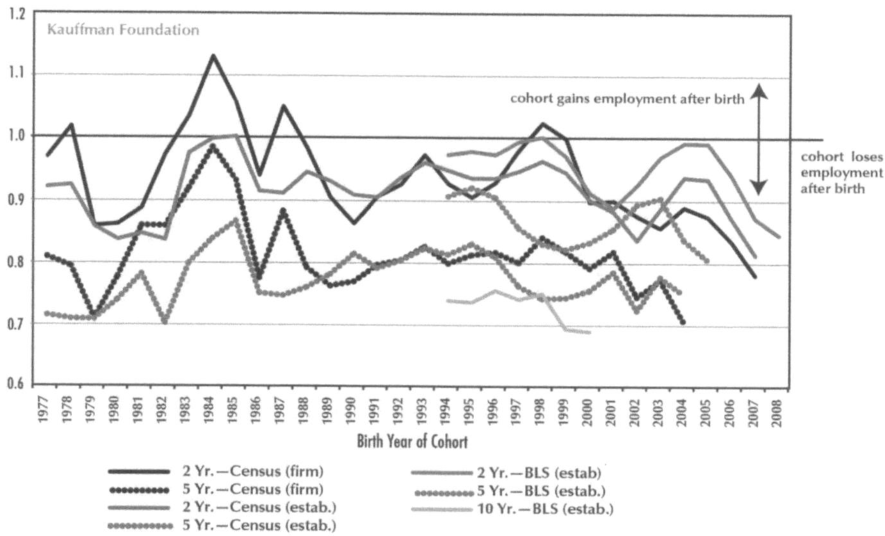

Fig. 11 Total employment of cohort as percentage of base-year employment (establishments and firms) (Source: Author Calculations from Census Bureau's Business Dynamics Series (published series and unpublished establishment portions of the series) and Bureau of Labor Statistics' Business Employment Dynamics Program)

additions has been trending downward for business cohorts since 1994. The average rate of employment growth from birth to age 2 and then age 2 to age 5 has been decreasing in all the data series, with only moderate yearly variation. So, while the levels might vary slightly in the different data series, the trends appear similar: *Businesses that survive their early years of existence have been adding jobs at a slower pace than the historic norm in recent years.*

So far, we have examined data relating to the size of the *average business*. What about *aggregate employment of new business cohorts* over time? Here we compare total employment of the cohort's surviving members over time to the baseline (birth) level of employment (Fig. 11).

Both the BLS and Census data show a similar pattern. The business cohorts continue to shed more jobs in the aggregate than they add in total over the years (thus staying below 1 on Fig. 11). This means expanding businesses did not compensate by adding employees at a fast enough pace for the business cohort to hold onto the total employment it started with once jobs at exiting businesses were accounted for.

Figure 12 shows that, when measured at age 2, cohorts of new firms born after 2001 retained about 90% of their total employees. For the cohort born in 2007, the retention rate had dropped to less than 80%. Those same cohorts also retained fewer jobs at age 5, down from the historical average of 80% prior to the 2001 cohort to slightly more than 70% in the most recent cohort (2004 firms). *That ten percentage-point decline in employment retention is equivalent to about 300,000 fewer jobs per business cohort.*

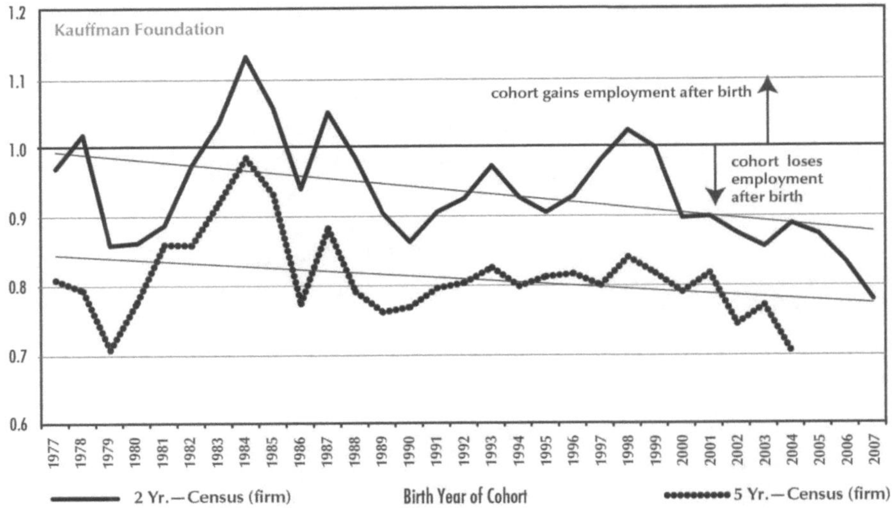

Fig. 12 Total employment of cohort as percentage of base-year employment (Census firms) (Source: Author Calculations from Census Bureau's Business Dynamics Series (published series and unpublished establishment portions of the series) and Bureau of Labor Statistics' Business Employment Dynamics Program)

The degree of decline as establishments grow older differs somewhat between the Census and BLS series. Census data show establishment cohorts shedding jobs more quickly (Fig. 13) in the aggregate than BLS does (Fig. 14). BLS shows establishments starting smaller, but holding onto jobs better. Census shows establishments starting larger, but shedding jobs faster. In the end, however, both series show a decidedly negative jobs growth picture in recent years.

Summing up, one key undisputed finding emerges: *The employment that a new business cohort is born with in the United States is likely the maximum employment that business cohort will experience in its lifetime.*[13] That has been true even when new businesses were generating a greater number of jobs than they are now.

When we see shrinking trends in employment from new businesses generally, as shown earlier in the report, there is no evidence from the data series we have examined that the *aggregate* contribution of jobs from that business cohort will ever be anything except smaller, even though individual businesses may scale substantially. So, for example, for firms born in 2009, the 2.3 million jobs they created – a

[13] This finding is consistent with the data examined by Horrell and Litan (2010), showing that startup cohorts tended to keep about 80% of the jobs they initially created (though distributed differently among firms). Additionally, it should be noted, although not presented here, this trend toward early maximum employment contribution holds true at the most common broad industry categories available in the data. Thus, while there is certainly some part of the story here which is attributable to changes in the industries in which new businesses are being formed, this particular trend appears true.

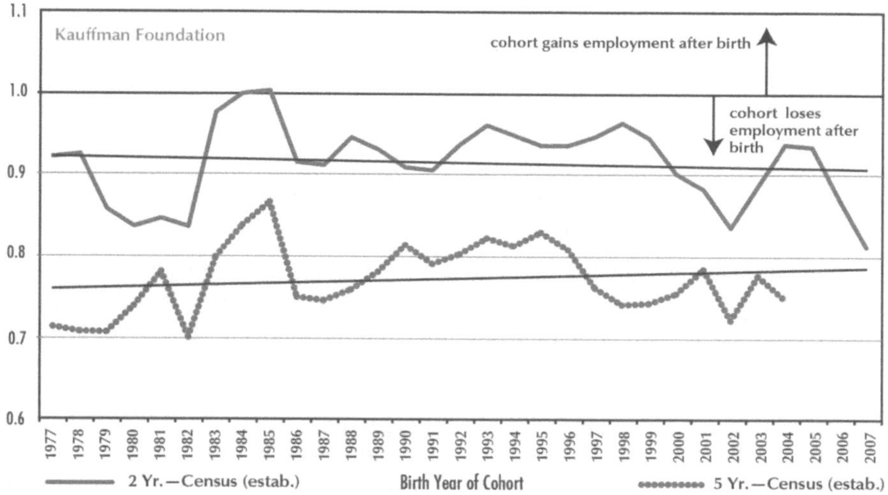

Fig. 13 Total employment of cohort as percentage of base-year employment (Census establishments) (Source: Author Calculations from Census Bureau's Business Dynamics Series (published series and unpublished establishment portions of the series) and Bureau of Labor Statistics' Business Employment Dynamics Program)

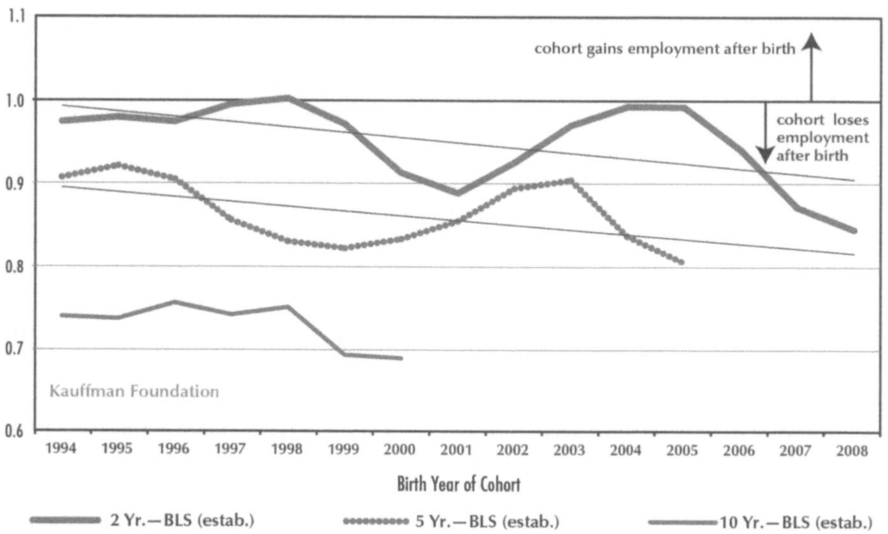

Fig. 14 Total employment of cohort as percentage of base-year employment (BLS establishments) (Source: Author Calculations from Census Bureau's Business Dynamics Series (published series and unpublished establishment portions of the series) and Bureau of Labor Statistics' Business Employment Dynamics Program)

figure considerably below the recent historic norm – are likely to represent the peak level of this cohort's contribution to U.S. employment. The relatively restrained job creation of new businesses in recent years will, therefore, have a substantial negative impact on longer-term employment levels.

6 Limitations of This Research and Looking Ahead

It is important that we qualify our findings by noting the limitations of the available data. In this essay, we have presented trends only at the aggregate levels within the United States. There are limited data of the type we studied available for study in Europe, but we have not analyzed those yet and they are not strictly comparable across countries. Very few other countries have the capabilities to track business dynamics sufficiently to capture some of these trends.

Additionally, we are concerned about the divergence between the BLS and Census data in regard to employment, in particular, and urge researchers in the future to examine this issue. Indeed, without further exploration and improvements in these sources, it will be difficult to unpack the underlying drivers in some of the trends that could be related to increased firm-level productivity, shifting occupational/employment needs at the firm level resulting from information technology, globalization, and other drivers, or, more likely, some combination of all of the above factors and others not yet recognized.

Finally, it should be noted that we were only able to examine the aggregate and average data in the datasets, which might not be telling the whole story. A fuller examination of the micro-data would look at shifts in the full distribution of entry and growth patterns over time to see if the jobs problem comes more from shrinking numbers of scale entrants or a relatively even downward trend across all firms. While we characterize the process uniformly as "up or out," there is actually a lot of variation at the firm level over time that merits further study.

Nonetheless, the conclusions from the data analyzed here are pretty clear, and they are not heartening. Employer businesses have been starting in fewer numbers, with fewer employees, growing slower, and, therefore, generating increasingly fewer new jobs for the U.S. job market. Furthermore, cohorts of businesses are just like cohorts of people. Once they are born, many of their characteristics are fairly well imprinted. In the case of the United States, business cohorts' peak employment contributions almost always occur in their birth year. Thus, the falling contribution of new businesses to job creation will be felt for years. If these business cohorts follow historical trends, we can expect them to contribute fewer jobs for at least their first decade. Already recent cohorts of new businesses have experienced historical lows in their ability to retain jobs. While these trends are not written in stone, changing these patterns will not be simple.

At a time when policymakers are focused on finding solutions to sluggish employment growth, the role of young firms and establishments on broader employment growth is poorly understood and appreciated. Two implicit assumptions in the debate about jobs, which we believe are wrong, deserve mention.

First, policymakers' focus on big changes in employment because of events such as a new manufacturing plant or the recruitment of a business to a community ignore the more important fact that our jobs outlook will be driven more by the collective decisions of the millions of young and small businesses whose changing employment patterns are not as easy to see or influence.

Second, it is just as easy to be deluded into thinking that the jobs problem will be solved by growth in the number of the self-employed. The last decade has brought about some fundamental shifts in U.S. labor markets, including the rise of outsourcing, not just to foreign locations.[14] In many cases, companies or individuals that once would have been hired as employees of a business now are performing the work on a temporary basis as contractors through other professional service organizations or under their own self-employment contracts. These individuals, while sometimes characterized as "entrepreneurs," are not likely to employ others or to reach significant scale. No matter how laudable their individual efforts, these sole proprietors, almost by definition, are not likely ever to be major employers.[15]

The clear challenge for the U.S. economy instead is to start more *employer businesses, ensure that they are starting larger,* and *nurture their growth.*

References

Bravo Biosca A (2010) Growth dynamics exploring business growth and contraction in Europe and the US. November: NESTA

Davis SJ, Haltiwanger J, Jarmin RS, Krizan CJ, Miranda J, Nucci A, Sandusky K (2009) Measuring the dynamics of young and small businesses: integrating the employer and nonemployer universes, NBER chapters. In: Timothy Dunne, J. Bradford Jensen, Mark J. Roberts Producer dynamics: new evidence from micro data. National Bureau of Economic Research, pp 329–366

Haltiwanger JC, Jarmin RS, Miranda J (2010) Who creates jobs? Small vs. large vs. young, August: NBER working paper no. 16300

Horrell M, Litan R (2010) After inception: how enduring is job creation by startups? Kauffman Foundation Research Series: Firm Formation and Economic Growth, Kansas City, MO

Izzo P (2011) Nearly 1 in 3 unemployed out of work more than a year, June 3: wall street journal. http://blogs.wsj.com/economics/2011/06/03/nearly-1-in-3-unemployed-out-of-work-more-than-a-year/

Kowarski I (2011) Freelance jobs: half of all new jobs in recovery?, June 13. http://www.csmonitor.com/Business/2011/0613/Freelance-jobs-Half-of-all-new-jobs-in-recovery

Samuelson L (2011) How to avoid our own lost decade, June 12. http://www.ft.com/intl/cms/s/b3c143b6-952d-11e0-a648-00144feab49a

U.S. Bureau of Labor Statistics (2011) How long before the unemployed find jobs or quit looking?, Issues in Labor statistics, Washington, DC

[14] A recent article by Kowarski (2011) also highlights some of these issues.

[15] Indeed, in this study we have ignored self-employment patterns other than as included in the Kauffman Index of Entrepreneurial Activity. Earlier work, some of which Kauffman has sponsored, has attempted to better integrate data on the self-employed with the business data examined here (See Davis et al. (2009)). Although this study shows that a non-trivial percentage of employer businesses have histories, including a period of self-employment, larger-scale jobs growth for the economy as a whole is not likely to emerge from an increase in self-employment.

The Italian Small- and Medium-Sized Firms in the Aftermath of the Crisis

Giorgio Calcagnini and Ilario Favaretto

1 Introduction

The outbreak of the recent crisis that began at the end of the last decade eventually uncovered the problems that almost all Western economies had been able to camouflage thanks to economic growth fueled by an unprecedented expansion in the financial industry. Within this context the Italian economy is a special case: its annual growth rate has been lower than that of its traditional partners and competitors for the last 15 years. Even the most optimistic forecasts for the next several years do not indicate significant improvements in the Italian competitive position. The success it enjoyed following its participation in the European Monetary Union brought to the fore the structural backwardness of the Italian economy, which is the main reason behind its low growth rates. The complexity and costs investors face when starting a business, the complexity and inefficiency of the fiscal, credit, and the legal systems, the difficulties associated with investor protection are only some of the obstacles that entrepreneurs face at the time of investing in Italy. More recently, we need to add a new weakness, that – until a few years ago – was considered a key factor in its success, i.e., the size of Italian firms.

As already shown in previous work,[1] the Italian economy is characterized by the large presence of small- and very-small-sized firms which is not balanced by a significant presence of medium- and large-sized firms, as is the case for other economies (i.e., the U.S.). Indeed, during the two worst years of the crisis, 2008 and 2009, the Italian GDP growth rate was, on average, the lowest among a group of

[1] See Calcagnini and Favaretto (2011).

G. Calcagnini (✉) • I. Favaretto
Dipartimento di Economia, Societá, Politica, Università di Urbino "Carlo Bo", via Saffi 42, 61029 Urbino, Italy
e-mail: giorgio.calcagnini@uniurb.it; ilario.favaretto@uniurb.it

G. Calcagnini and I. Favaretto (eds.), *Small Businesses in the Aftermath of the Crisis*, Contributions to Economics, DOI 10.1007/978-3-7908-2852-8_5,

industrialized countries. Further, there seems to be a positive relationship between firm size and GDP growth rates.

This chapter aims at analyzing trends in the Italian production structure before and during the current economic crisis, and changes in the decentralization process following the most recent technological innovations and functional re-organization among firms. More specifically, it describes how the latter process has been slowly assigning an increasing role to better-structured firms with respect to the past.

Results show that the long-time quest for the firm size most suited to the Italian economy has recently taken a radical change of direction. The new competitive context favors those firms that more intensively invested in strategic variables such as innovative technologies, human capital, and complex organizations within which each individual firm coordinates and guarantees the production phases characterized by higher quality content. As a consequence, we can observe a transformation in the Italian production system from a situation characterized by a very large number of small- and very small-sized firms, low innovation levels, and a large share of irregular economy into a new system where firms progressively find investing in R&D, participating in inter-industry organizations, selecting more educated employees and providing better training to be rewarding. Furthermore, just as important would be for firms to increase their size to become more profitable. All of the aforesaid would allow firms to cope with stricter, more competitive markets at the international level. This transformation is necessary to allow Italian firms to keep competing internationally, even though it will likely come at the cost of a smaller production base (a lower number of firms). However, the change outlined would be in line with EU policies supporting regional and endogenous development projects that make the most of the above-mentioned strategic variables.

The chapter is organized as follows. The first two sections (Sects. 2 and 3) describe changes in the Italian production system up to the years before the current economic crisis, while Sects. 4, 5, 6, 7, and 8 show how firms are reacting to it. Sections 9 and 10 sketch the most important features of the new Italian production model, while Sect. 11 concludes.

2 Trends in the Italian Production Structure: From 1970s to 1990s

During the 1970s, technological development increased labor productivity in many industries but, at the same time, reduced the level of the "optimal" number of workers per given output. In other words, in many industries the diffusion of technological innovations allowed firms, even small-sized ones, to reduce their efficient scale of production. Thus, particularly small-sized firms managed to continue producing, despite the presence of less standardized demand, and expanding at lower growth rates than in the previous period. Further, this first wave of technological innovations enabled them to better utilize their production factors associated with the maximum elasticity in production organization. Therefore, between 1971 and 1981, the increase in the number of firms was larger than that of workers and, consequently, the average firm size decreased.

During the period 1981–2001, the decentralization of production became functional, and not substitutive with respect to the innovation and rationalization processes of the Italian economy. Further, the development of small-firm systems was mainly the result of the valorization of local production opportunities and was not driven by the restructuring needs of large-sized firms. The subsequent strengthening of the competitiveness of Italian production was the result of firms' increased flexibility, which followed their re-organization, higher production efficiency and, finally, the consolidation of systems of small-sized firms in local markets.

The increasing labor division between firms reflected their larger decision-making autonomy (which was confirmed by phenomena such as their multiple customer bases and their production diversification), and was not merely the result of small-sized firms' dependence on larger firms. Further, the production of investment goods that incorporate high levels of technology and the specialization of even very small-sized firms allowed the latter to compete internationally.

Between 1981 and 2001 a new wave of re-organization of the Italian production system determined a reduction both in the number of firms and workers. However, at the same time, some of the small-sized firms made directly sure that customers' order were filled to satisfaction and started strengthening their internationalization process. Indeed, two types of production delocalization were still at work. On one hand, we find firm systems within which technologies and innovations did not play a significant role in determining their competitiveness and the degree of market competition in which they operated (*cascading delocalization*). On the other, we find more developed firm systems within which technology played an increasing role; firms acquired specific investment goods (equipment and machinery) through changes made to standard investment even by the smallest firms. These changes occurred to meet the most diversified firm needs and were most often the result of entrepreneur and worker experience. They resulted in small, but systematic improvements in firm production processes and products. However, the ability to attain these improvements depended upon recruiting workers and training them and was significantly differentiated at the regional and industry level.

At the same time we can observe advanced decentralization processes (*functional complex relationships*) characterized by technological innovations. Here small-sized firms played an active role within the system of functional relationships among companies and kept up with changes in technology, organization and the markets in which they operated.[2]

In short, during the 1970s (1971–1981) we witness a decrease in the average firm size and a simultaneous, relatively larger increase in the number of firms rather than

[2] See (Favaretto 1995).

in the number of employees. Firm strategies were mainly directed at "slimming" production processes by outsourcing. Further, the degree of vertical integration decreased. On the contrary, during the 20 years that followed (1981–2001) decentralization processes occurred by means of heavy restructuring efforts that determined a contraction in both the number of firms and workers.

3 Signs of Economic Distress Precede the Crisis

The stop that occurred in the process of quantitative increase in the number of small firms between 1990 and 2003 went hand in hand with the slowdown of the Italian economy as measured by its GDP growth rate. On average, the Italian growth rate was 1.47%, significantly lower than that of the EU-15. Over the period, the cumulated difference in the growth rates of the two areas was 7.3% points. This gap is also confirmed when the per-capita GDP growth rates are taken into account; it is related to the structural characteristics of the Italian economy such as its production specialization in industries characterized by a medium-low technological intensity, which was moreover largely exposed to price competition from emerging countries. Further, value added and labor productivity in these industries was relatively low compared with other industries.[3]

The production structure is one of the main causes of the difficulties experienced by Italian economic development. When the decentralization process is analyzed by industry, using the Pavitt taxonomy, the negative contribution to the creation of new firms during the period 1981–2001 mainly comes from supplier-dominated industries.[4] The latter are dependent on external sources of innovation, mainly the suppliers of machinery and materials. The supplier-dominated industry includes firms from traditional manufacturing such as the textile industry.

Specifically, during the period 1971–2001, the size of firms belonging to supplier-dominated industries (textile, clothing, leather, shoes, wood and furniture – mostly small-sized firms) was relatively stable. In the same period, industrial districts

[3] The low growth rates of the Italian GDP depend on a stagnating productivity. The latter, between 1997 and 2007, increased on average by 0.4%, that is less than one-third of the EU-15 and one-fourth of the German one.

[4] The Pavitt (1984) taxonomy of industrial sectors categorizes industries according to determinants and patterns of technological change. It uses the nature and sources of innovation, technology user types and means of appropriability as central factors used in industry classification. The Pavitt taxonomy categorizes firms into four groups: (1) scale-intensive, (2) supplier-dominated, (3) specialized-supplier and (4) science-based.

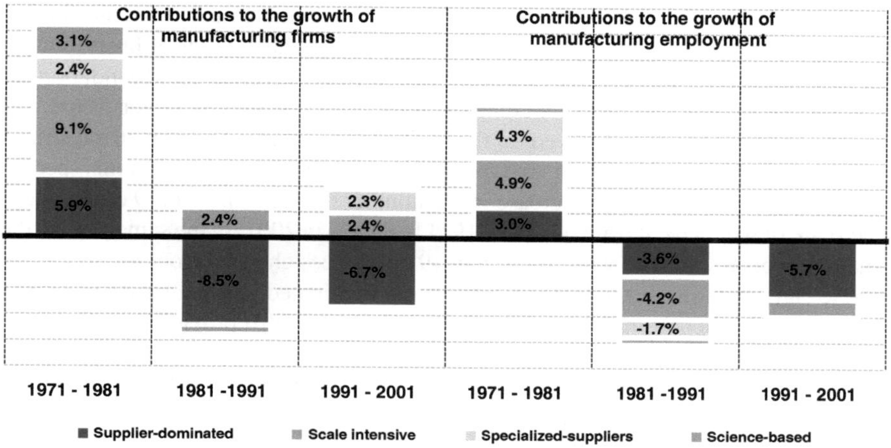

Fig. 1 Contribution to the growth of Italian firms and employment by industries classified according to the Pavitt taxonomy (1971–2001)

experienced a strengthening of their production model, especially within supplier-dominated industries.[5]

During the same period the Italian economy was hit by several economic shocks that allowed the production model based on small-sized firms to gain ground, thanks to its capacity to quickly and better adjust to changing market conditions.

Within scale-intensive industries (steel, glass, durable goods such as automobiles, etc.), where it is easier to find large-sized firms, average firm size progressively decreased during the period 1971–2001.

Furthermore, firms belonging to specialized-supplier industries, which typically include investment goods industries and large-sized firms, experienced a decrease in their average size. However, the largest decrease in average firm size was observed in science-based industries (electronics, pharmaceuticals, electric machineries, etc.) (see Fig. 1).

During the last two decades, the average firm size in supplier-dominated industries showed a small increase, following a decrease in the number of workers, which was proportionally smaller than the decrease in the number of firms.

[5] By the term 'industrial district' we mean a socio – territorial entity characterized by the active co-presence, in a confined area with natural and historical properties, of a community of people and a population of industrial firms (Becattini 1989). Accordingly to its orthodox definition, the industrial district model is based on the presence of only one type of industry, together with a set of economies external to firms, but internal to the industry. These economies came out of the deconstruction of the production process (and in a corresponding manner the division of labor) within each industry (Fadda 2010). However, the industrial district model does not consider that today significant and increasing advantages are possible thanks to the positive externalities coming from the co-presence, within an industrial district, of more industries and the economies linked to material and immaterial public goods connected to a specific area (Favaretto 2003).

As for the specialized-suppliers and scale-intensive industries, their contribution during the last decade mainly concerned the growth in the number of firms, but not in employment. Indeed, their outsourcing strategy favored the foundation of specialized small-sized firms that employed the workers fired by larger firms.

Trends similar to that observed for the last two types of industry also characterized science-based industries, which are typically not very labor intensive. In this specific case there was a remarkable increase in the number of micro firms (1–9 employees), the share of which went from 67.7% in 1971 to 87.4% in 2001. In the same period, the share of workers in firms with fewer than 19 employees changed from 4% to 23.7%. These figures show that relatively large firms dominated science-based industries in terms of job opportunities but, at the same time, they also provided smaller firms with many opportunities to start new enterprises.

It is once again likely true that the key to interpreting what happened in the period that followed the crisis can be found in the complex dynamics of the decentralization processes of Italian industry.

4 The Decade of the Crisis: The Decrease in the Vertical Integration of Production and Trends in Average Firm Size

During the period 2002–2007, by using the Pavitt taxonomy once again, it is possible to determine that the index of vertical integration showed a generalized decrease in scale-intensive, specialized-supplier and science-based industries. Only for supplier-dominated industries does the index show a significant volatility around a relatively stable mean value (see Fig. 2).

During the same period, within supplier-dominated industries, small-sized firms increased their employment share, even though average firm size decreased slightly from 6.1 to 5.8 employees.

On the contrary, scale-intensive industries increased their average size from 9.7 to 10.1 employees, mainly as a result of the decrease in the number of firms.

The average size of firms in specialized-supplier industries was essentially stable around 13.5 employees, but these companies' share decreased with respect to the other types of industries.

Finally, science-based industries showed a reduction in the number of firms, but compensated through a consolidation of their average firm size (from 11.3 to 11.8 employees) and in their employee share (see Fig. 3).

At the beginning of the last decade, with the exception of the supplier-dominated industries, the Italian manufacturing industry experienced a new process of production decentralization. The most recent form has been organized differently from the previous one. It is no longer based on the continuous increase in the number of flexible small-sized and micro firms, which allowed the Italian economy to face and survive the consequences of recurrent business cycles. The new decentralization process is instead centered on more advanced technologies, a more modern

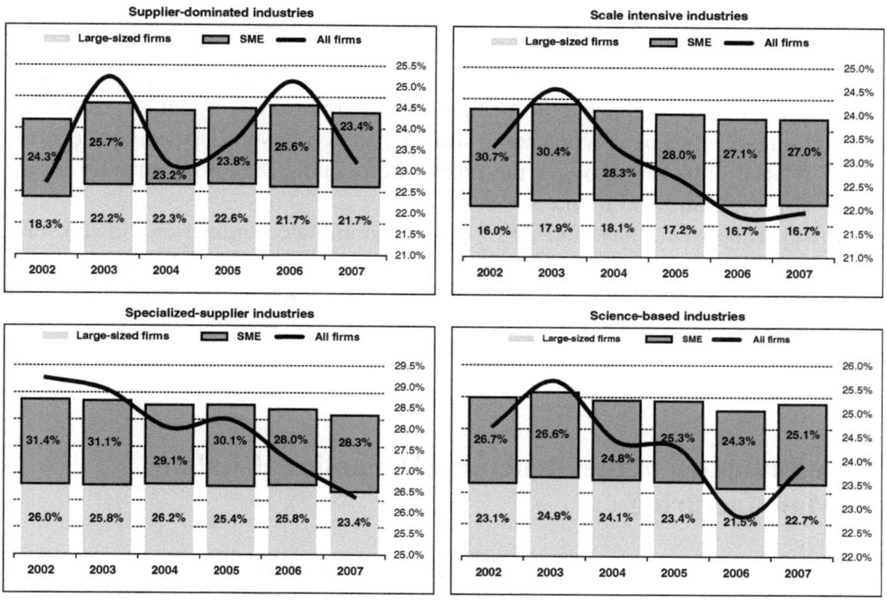

Fig. 2 Index of vertical integration for manufacturing industries classified according to the Pavitt taxonomy (2002–2007)

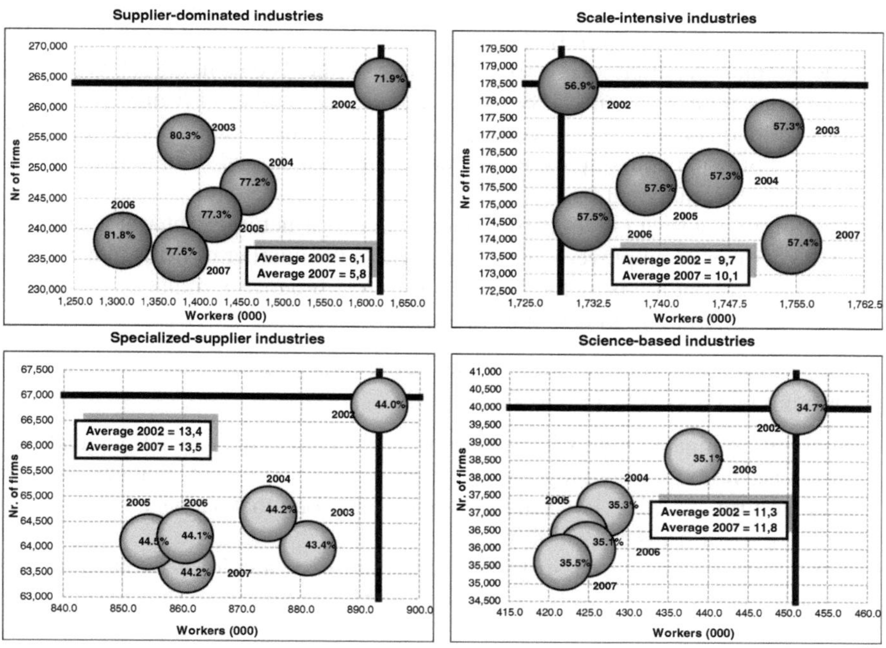

Fig. 3 Changes (%) in the number of firms and employment of manufacturing industries classified according to the Pavitt taxonomy (2002–2007)

business organization or an initial product differentiation, all of which are the right instruments to face the competitive challenges that go along with globalization.

On the contrary, supplier-dominated industries, which characterize the Italian economy, are still organized according to the traditional model in which the production decisions of subcontractor firms depend on buyers' operative decisions. These relationships are stronger the more often small-sized firms operate with a single buyer and in local markets.

However, between 2000 and 2007, supplier-dominated industries showed some changes in their structural characteristics; therefore, a more detailed investigation should be carried out before stating that supplier-dominated industries have been fundamentally static.

5 A Summary of the Structural Dynamics of Italian Manufacturing

The time evolution of industries' average size depends on two variables, i.e., the growth rates of employees and the number of firms. It follows that, on a Cartesian graph, we can locate four clockwise quadrants corresponding to the different evolutionary phases of the Italian production structure (see Fig. 4). Therefore:

- The first quadrant (+, +) corresponds to the "growth" phase: an increasing number of firms and employees;
- The second quadrant (+, −) corresponds to the "consolidation" phase: an increasing number of employees and a decreasing number of firms;
- The third quadrant (−, −) corresponds to the "stagnation" phase: both the number of firms and employees decrease. First comer and old industries often belong to this quadrant.
- The fourth quadrant (−, +) corresponds to the "restructuring" phase: a decrease in the number of employees and an increase in the number of firms that react to new market conditions by looking for a different production-input combination, often by undergoing process innovations.

Results from this analysis show marked differences among industries classified according to the Pavitt taxonomy (see Fig. 4).

- Supplier-dominated industries are almost all located in the stagnation quadrant: they lose firms and employees and there are no significant changes in their average firm size. The Food industry is the only one that shows an increase in the number of firms and employees.
- Most firms belonging to scale-intensive industries are located either in the growth quadrant (firms with 20–49 and 50–250 employees) or the consolidation

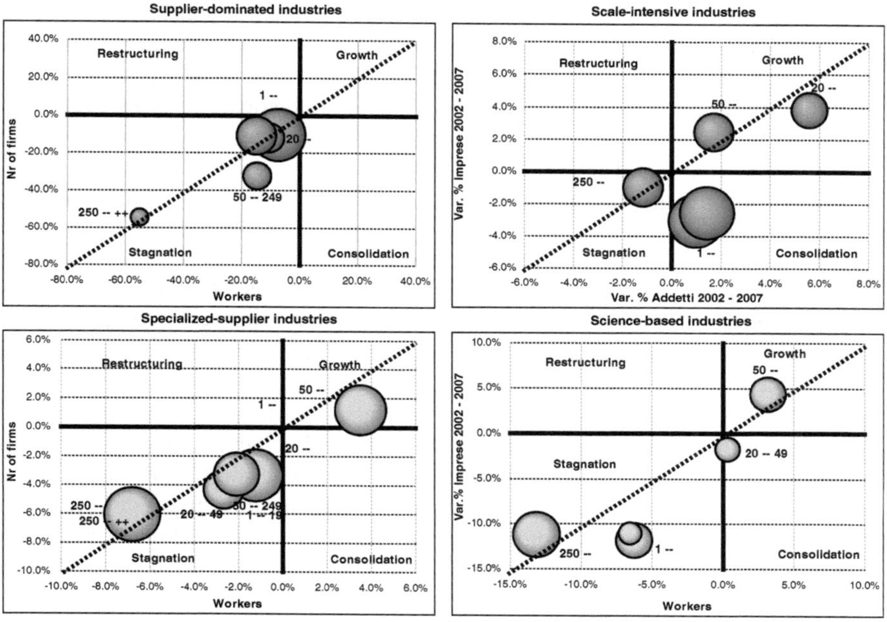

Fig. 4 "Evolution phases" of the Italian manufacturing industries classified according to the Pavitt taxonomy (2002–2007)

quadrant (firms with 1–19 employees). Only larger firms, i.e. those with more than 250 employees belong to the stagnation quadrant.

- Specialized-supplier industries are mostly located in the stagnation quadrant, with firms experiencing a decrease in their average size. Firms with 50–250 employees represent the only exception and they are located in the growth quadrant mainly thanks to the investment goods and the other transport goods industries.
- The picture for science-based industries is more complex than it is for previous industries. Large- (more than 250 employees) and small-sized (1–19 employees) firms are located in the stagnation quadrant, while firms with 20–49 employees show signs of having undertaken a consolidation process. Finally, only firms with 50–250 employees belong to the growth quadrant.

Briefly, during the years of the last decade that preceded the current crisis, the growth of manufacturing dominated by micro- and small-sized firms, belonging to supplier-dominated industries and characterized by "cascading" production delocalization, came to a stop.[6]

[6] See Favaretto (1995).

Data referring to the most recent period allow us to understand trends, and the intensity that characterizes changes in the Italian economy and in other economies. In 2008 and 2009 firm entry growth rates significantly decreased in all the most important European economies and the U.S.A. simultaneously with an increase in exit growth rates. Therefore, firm net growth rates decreased, but remained positive except for Italy and the U.S.A. (see Table 1).

In Italy firm net growth rates have been decreasing since 2004[7] and in 2007 the exit growth rate recorded its highest value for the period 2003–2009. In 2009 the total number of artisan firms was lower than in 2008 (see Table 2).[8]

The system of Italian small- and medium-sized firms (SMEs) is facing a crisis that, on one hand, sanctions the end of the traditional *expansive* model and, on the other, favors a new model based on the search for higher qualitative standards in firm production and organization.

The transition from one model to the other is also affecting the Italian position on international markets. Indeed, traditional competitive advantages of supplier-dominated and scale-intensive industries have been deteriorating during the last decade relative to science-based and specialized-supplier industries.[9]

Data on Italian exports and imports (in current values) show that the export share of goods embodying high intensity R&D expenses systematically decreased from 10.7% in 2001 to 9.8% in 2005, falling to 9.3% in 2010 (see Fig. 5).

As for imports, the share for the same type of goods increased from 17.2% in 2001 to 18.2% in 2010 (it was 16.6% in 2005) (see Fig. 6).

The export share of supplier-dominated industries also decreased between 2001 and 2010, from 35.3% to 30.0% (it was 31.6% in 2005), while imports increased slightly from 24.4% to 25.5% as a result of a rise in international competition (see Figs. 5 and 6).

Exports of specialized-supplier industries increased from 22.6% to 24.9%, while imports decreased from 14.3% to 12.8%. The same type of trend occurred in scale-intensive industries, which witnessed an increase in exports (from 31.4% to 35.8%) and a decrease in imports (from 44.0% to 43.5%) (see Figs. 5 and 6).

Products more frequently supplied by small-sized firms, i.e. those belonging to supplier-dominated and specialized-supplier industries, modified their weight within the category of Italian international trade between 2001 and 2010. Specifically, the lower weight of exports from supplier-dominated industries and the corresponding increase in those from specialized-supplier industries reflects a larger contribution of small-sized firms to the demand of high-intensity technology products. On the other

[7] According to ISTAT (2010a) the turning point occurs in 2003.

[8] Data shown in Tables 1 and 2 come from different sources. This explains the differences in Italian growth rates in the two tables. However, both data sources show a consistent time decline in Italian growth rates, which is the most important result we want to focus on.

[9] See Rombaldoni's chapter in this volume.

Table 1 Firm growth rates for a sample of countries (2006–2009)

	Belgium	Finland	France	Germany	Ireland	Italy	Netherlands	Spain	UK	Japan	US
2006	2.1	3.6	1.4	1.3	6.9	0.9	2.6	2.5	2.8	−1.8	0.8
2007	3.1	4.0	2.1	0.6	6.3	0.0	3.3	3.5	4.0	–	0.5
2008	2.3	2.3	1.4	0.2	–	−0.8	4.1	1.4	0.6	–	−0.6
2009	0.7	–	0.4	0.6	–	−2.6	2.1	2.6	0.5	–	−2.2

Source: European Commission "European SMEs under Pressure", (European Commission 2010)

Table 2 Italian firm demography: all firms and artisan firms (2003–2009)

	Entry rate	Exit rate	Growth rate
2003	6.68	5.23	1.45
2004	7.21	5.43	1.78
2005	7.02	5.41	1.61
2006	6.97	5.77	1.21
2007	7.12	6.37	0.75
2008	6.71	6.11	0.59
2009	6.32	6.03	0.28
Of which: artisan firms			
2003	7.95	6.77	1.17
2004	8.65	7.30	1.35
2005	8.30	7.26	1.04
2006	8.22	7.51	0.71
2007	9.25	8.41	0.84
2008	8.40	8.03	0.37
2009	7.25	8.32	−1.06

Source: Calculations based on Unioncamere-Infocamere (Movimprese) data

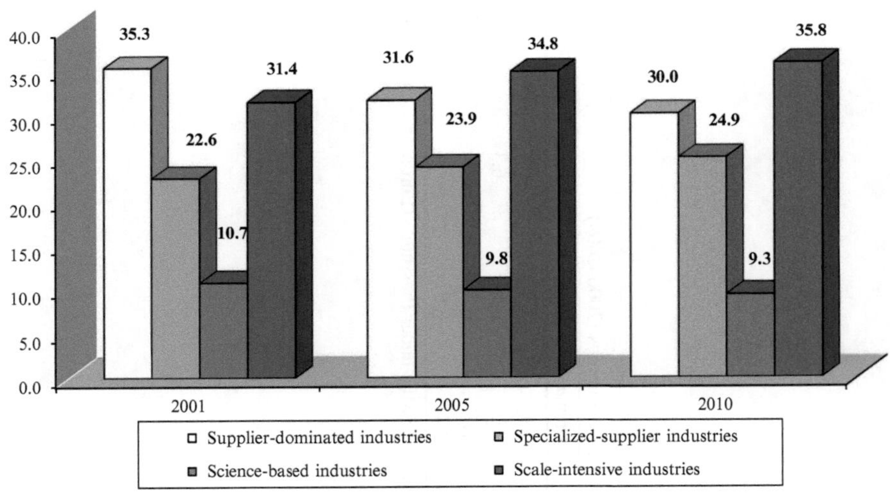

Fig. 5 Italian manufacturing export shares by industry classified according to the Pavitt taxonomy (2001–2005-2010) (Source: Calculations based on ISTAT-COEWEB data)

hand, changes that occurred in imports (an increase in the case of supplier-dominated industries and a decrease for specialized-supplier industries) show that small-sized firms suffered from international competition inversely with the technological intensity of their products.

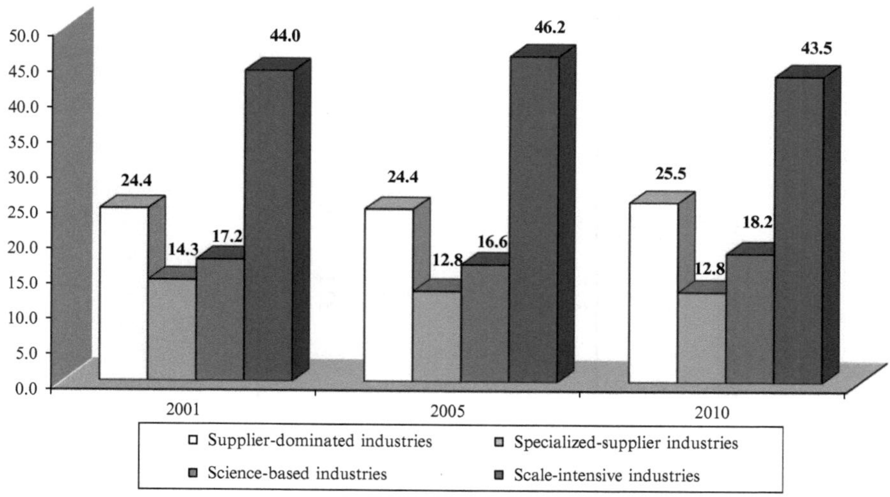

Fig. 6 Italian manufacturing import shares by industry classified according to the Pavitt taxonomy (2001–2005–2010). Source: Calculations based on ISTAT-COEWEB data, ISTAT (2011)

6 How Italian Firms That Restructured and Invested in R&D Are Reacting to the Crisis

A report from the Banca d'Italia (2010) showed that, between 2000 and 2006, 54% of Italian firms undertook a restructuring process and in the following 2 years (2006–2008) 55.1% made R&D investments.[10] The report shows that restructured firms were hit by the crisis less than firms which did not undertake the restructuring, and their entrepreneurs have, on average, more optimistic growth expectations. These results hold for most of the Italian regions, with the exception of those located in the Mezzogiorno (i.e. the southern Italian regions), which made lower restructuring and R&D investments than the regions in the North and the Centre.

However, even firms that restructured and invested in R&D reported negative changes in their sales during 2009, but their share was almost everywhere lower than the share of firms that carried out no restructuring. Similarly, the relative number of firms that expected to increase their sales is larger among those that restructured and invested in R&D (see Tables 3 and 4).

The Banca d'Italia report also shows that many firms faced the crisis by changing, i.e. introducing new products, by making investments in trademarks and by internationalizing. Some firms, especially those located in the North-Eastern

[10] The restructuring was forced by the increase in international competition and by new opportunities created by the concomitance of technological revolution, globalization, and European integration by means of the adoption of a single currency (euro) at the end of the 1990s. The restructuring process was characterized by product and process innovations, and the increase in the degree of internationalization of Italian firms. See Brandolini et al. (2010).

Table 3 Trends in sales and investment, and the restructuring of Italian manufacturing firms (weighted averages, shares, and number of firms)

	North-West		North-East		Centre		South		Italy	
	Restructured firmsin 2000–2006[a]	Other firms	Restructured firmsin 2000–2006[a]	Other firms	Restructured firmsin 2000–2006[a]	Other firms	Restructured firmsin 2000–2006[a]	Other firms	Restructured firmsin 2000–2006[a]	Other firms
Nr. of firms	226	155	172	144	204	152	297	322	899	773
Share (%)	59.3	40.7	54.4	45.6	57.3	42.7	48.0	52.0	53.8	46.2
Decreasing sales in 2009[b]	71.0	81.9	70.6	68.4	72.9	80.1	63.0	67.5	70.2	74.5
Decreasing investment in 2009[b]	64.9	73.0	60.9	56.5	63.0	70.9	69.6	66.3	63.9	65.7
Increasing sales in 2010[b, c]	50.5	43.0	51.4	41.2	48.9	44.0	41.4	47.7	49.3	43.4
Increasing investment in 2010[b, c]	44.6	38.5	41.0	39.3	42.2	33.6	31.6	32.1	41.4	36.8

Source: Banca d'Italia (2010)
[a]Estimated values referred to those firms interviewed in 2006 on actions taken in 2000–2006
[b]Percentage of positive answers
[c]Forecasted positive answers obtained in March–April 2010

Table 4 Trends in sales and investment, and the Italian manufacturing R&D investment (weighted averages, shares, and number of firms)

	North-West		North-East		Centre		South		Italy	
	Firms that invested in R&D in 2006–2008 (1)	Other firms	Firms that invested in R&D in 2006–2008 (1)	Other firms	Firms that invested in R&D in 2006–2008 (1)	Other firms	Firms that invested in R&D in 2006–2008 (1)	Other firms	Firms that invested in R&D in 2006–2008 (1)	Other firms
Nr. of firms	144	110	134	75	147	70	111	182	536	437
Share (%)	56.7	43.3	64.1	35.9	67.7	32.3	37.9	62.1	55.1	44.9
Decreasing sales in 2009 (2)	76.4	84.6	76.3	79.5	78.0	80.7	64.8	61.8	75.8	78.6
Decreasing investment in 2009 (2)	73.6	62.1	65.5	63.3	66.1	54.6	61.3	62.7	68.2	61.6
Increasing sales in 2010 (2) (2) (3)	53.6	48.3	57.1	30.2	53.0	53.4	60.2	47.8	55.3	43.7
Increasing investment in 2010 (2) (3)	48.0	44.5	48.9	28.1	41.3	38.0	34.4	35.5	46.0	37.3

Source: (Banca d'Italia 2010). (1) Estimated values referred to those firms interviewed in 2008 on actions taken in 2006–2008. (2) Percentage of positive answers. (3) Forecasted positive answers obtained in March–April 2010 regions, started internalizing production phases previously outsourced. During the last two years almost 50% of the surveyed industrial firms modified the type and quality of their products. This occurred more frequently among firms located in Central and North-Eastern regions. Further, firms kept searching for new markets, including by means of technical and production agreements with foreign firms. The share of firms that took over their competitors or suppliers remained, instead, relatively low.[11]

Some of the restructuring and rationalization processes aim at improving product quality and modifying the production organization by means of internalizing production phases previously outsourced. Eventually, the strengthening of these processes will significantly influence the structure of the Italian manufacturing industry and will lead to changes in the role and the numerical presence of micro-firms. For the time being, it is not completely clear if these changes are short-term or medium- long-term, but they necessarily require more in-depth analysis.[12]

7 Who Is Surviving and Who Is Better Off

To improve our understanding of firm characteristics that allow them to face the economic crisis in an efficacious manner, we focus on some performance indicators and their relationship with the average firm size (as measured by the number of employees) and age.

First, as illustrated in Calcagnini and Favaretto (2011, pp XII–XIII and Fig. 1), a positive correlation seems to exist between the GDP growth rate and average firm size. This relationship is measured during the 2-year period 2008–2009 and takes into account some institutional differences, such as those concerning labor and capital markets, among the countries included in the sample. Therefore, empirical results show that countries where firms are, on average, relatively smaller were hit more severely by the economic recession than countries with larger firms. The Italian economy, where firms are of the smallest recorded size, suffered the largest decrease in its GDP growth rate.

The existence of this positive relationship between firm size and GDP growth rates is also confirmed in other empirical analyses carried out at the regional level. A survey on recent sales trends for a panel of dynamic and innovative firms (also defined as "excellent" companies), located in the Italian Emilia Romagna region

[11] Banca d'Italia (2010), p. 5.

[12] The need to better understand the role of SMEs linked to outsourcing processes is well stressed by Accetturo et al. (2011, pp. 9–10). They recognize the existence of empirical evidence according to which, on one hand, SMEs play a marginal role within the Italian manufacturing industry, have low productivity levels and capital intensity. On the other, the authors show empirical results according to which SMEs are a heterogeneous aggregate where supplier-dominated firms coexist with innovating firms, in firms oriented to foreign markets, and in those with high productivity levels.

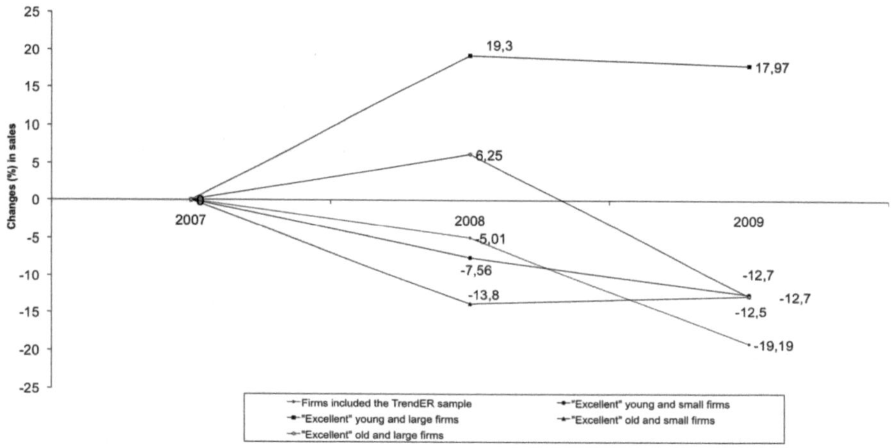

Fig. 7 Changes (%) in sales 2007–2009 for a panel of firms located in Emilia Romagna TrendER (2009).

and broken down by size and number of years from their foundation ("age"), shows a general sales decrease in 2008 and, especially, in 2009 (Fig. 7). However, some differences among the groups are worthy of note. Younger firms (with less than 15 years from their foundation) with more than 20 employees showed the best sales performances during the 2007–2009 period. Until 2008, older firms and those with more than 20 employees also registered an increase in sales. This result shows that more structured firms were capable of quicker reactions during the first phase of the crisis.

During the 2008–2009 period, instead, only younger firms with more than 20 employees showed a positive change in sales, while all other groups registered negative changes (Fig. 7).

This group of firms, as opposed to other groups, saw the crisis as an opportunity to redefine their competitive position and their market shares, especially internationally. Being younger and larger firms they are more apt to adopt new technologies and to closely follow the changes that occur at a fast pace. However, in order to fully exploit the advantages associated with the adoption of new technologies, these firms need to have an endowment of human, financial and organizational resources that, in most cases, only firms over a minimum size threshold (changing over time and from industry to industry) are able to guarantee.

8 The Consolidation of Italian Industry

The consolidation of Italian industry that has been taking place by means of an increase in firms' average size and the relative share of limited liability companies, can also be analyzed by the joint use of information on the dynamics of the

Table 5 Industrial Production Index (IPI) and Capacity Utilization (CU) (Averages of quarterly percentage changes)

(A) Business cycle	(B) IPI	(C) CU	(D = C/B) Elasticity	(E) CU Small firms	(F) CU Medium firms	(G) CU Large firms
'65-1quarter–'71-4quarter	0.90	−0.21	−0.23			
'71-4quarter–'75-2quarter	0.39	−0.70	−1.80			
'75-2quarter–'77-4quarter	1.20	0.63	0.52			
'77-4quarter–'83-1quarter	0.44	−0.17	−0.39			
'83-1quarter–'93-3quarter	0.39	0.07	0.19			−0.34
'93-3quarter–'96-4quarter	0.78	0.28	0.36			0.25
'96-4quarter–'03-2quarter	0.18	0.08	0.44	−0.18	−0.11	0.19
'03-2quarter–'09-2quarter	−0.77	−0.66	0.87	−0.65	−0.64	−0.61
'09-2quarter	1.57	1.75	1.11	1.70	2.29	1.50

Source: Calculations based on ISTAT, ISAE and ISCO data

industrial production index (that excludes the construction industry) and the degree of capacity utilization. The hypothesis is that the progressive re-internalization of production and the firm consolidation process may result in a more efficient use of firms' capital stock and, consequently, in relatively more intensive capacity utilization. To test this hypothesis we used quarterly data spanning from the fourth quarter of 1968 to the fourth quarter of 2010. Both series are seasonally adjusted by means of the X-12 ARIMA method.[13] The whole period was sub-divided into shorter periods according to the business cycle chronology provided for Italy by ISAE/ISTAT.[14] Troughs-to-troughs in economic activity measures the length of each business cycle (see Table 5).[15]

The time period between the end of the 1960s and mid-1970s, including the first two business cycles, is characterized by a strong increase in fixed investment and,

[13] The X-12 ARIMA method was firstly applied to the monthly industrial production index, and then quarterly averages of monthly data were calculated.

[14] Recently, the Istituto di Studi e Analisi Economica (ISAE), which had previously incorporated the Istituto per lo Studio della Congiuntura (ISCO), has itself been incorporated into the Italian National Institute of Statistics (ISTAT). In the text, we will refer to data originally provided by ISAE as ISAE/ISTAT.

[15] Table 5 shows the first quarter of 1965 as the beginning of the first business cycle, even though we only have data on capacity utilization starting from the last quarter of 1968. Therefore, calculations shown refer to the period 1968.4-1971.4.

consequently, in firms' capital stock. Therefore, it not unexpected that capacity utilization showed lower levels than in previous periods, notwithstanding the positive average growth rate in industrial production.

The next business cycle, from the second quarter of 1975 to the fourth quarter of 1977, records the first wave of industrial rationalization oriented to increasing efficiency. The rationalization process resulted in capacity utilization levels higher than those observed at the end of the previous business cycle (lower than 70%). During the same period, the average elasticity of capacity utilization with respect to industrial production was positive for the first time.

The business cycle that begins in the fourth quarter of 1977 and spans over the first quarter of 1983 is characterized by a positive industrial production growth rate, but a negative growth rate for capacity utilization. These are the years of the delocalization process of industrial production and the "discovery" of small businesses. Even though the Schumacher book was published in Schumacher (1973), "small is beautiful" remains the most popular slogan in academics, industry associations and politics. The production delocalization from larger to smaller businesses results in fewer advantages associated with scale economies in favor of more flexible organizations and production structures. This transformation is also possible thanks to the concentration of Italian firms in industries characterized by relatively low levels of capital intensity.

During the business cycle that began in the early 1980s and ended in 1993, the production model (and its industry specialization) that had prevailed in the previous period gained ground, to the progressive disadvantage of large-sized firms. During this period, on average, the utilization of capacity is relatively stable (but in the case of large-sized firms it decreases, on average, by 0.4% point per quarter), while industrial production increases, on average, by 0.4% per quarter. The latter is a value not too far from that observed during the two most critical business cycles of the 1970s.

From 1993 onward the Italian economy experienced four business cycles. This period is firstly characterized by the prospective participation of Italy in the European Union (EU) and, consequently, by a stricter monetary policy stance. Indeed, the success of the Italian economy until the beginning of the 1990s was also its ability to quickly adapt to changes in international markets, thanks to exchange rate movements that compensated for losses in the competitiveness of (small-sized) businesses. For a long period of time, the opportunity to use the exchange rate policy as the main intervention to support Italian firms internationally, along with the almost total lack of an industrial policy, made the defense of the traditional Italian position in international trade viable. During this time period no serious rethinking occurred on what, how and for whom to produce, while finding answers thereto was put off until the future. Therefore, from 1998, when Italy became one of the countries adopting the Euro, Italian industry found itself at a disadvantage compared to emerging economies, which became more and more competitive in those production processes for which the Italian economy had heretofore been more specialized. Further, the small size of Italian firms made defending their market shares even more troublesome.

Fortunately, something is changing. Notwithstanding the Italian institutional framework that does not favor "doing business"[16] nor encourage innovation, compared to other economies, firms are progressively responding to incentives and opportunities provided by the new international competitive context. One of these changes is the increase in average firm size; Italian average firm size has become more pronounced during the recent economic crisis, during which more scholars have reported a progressive re-consolidation within a single company of functions and operations that previously had been outsourced.

In our data this re-consolidation process may be observed in the increase in the elasticity of capacity utilization with respect to production. In the period between the business cycle that started in 1983 and today, the elasticity value almost tripled, showing that firms are using their production capacity more efficiently (Table 5).

It is only since the mid-1990s that ISAE/ISTAT have begun to provide data on capacity utilization broken down by firm size. Overall, during the last two business cycles, small-sized firms decreased their capacity utilization by 0.18% and 0.65%, respectively (Table 5). In a context of weak fixed investment growth, these changes in capacity utilization are a clear sign that production is progressively becoming more concentrated in larger firms than in small-sized ones.

The result is confirmed by data on mid- and large-sized firms. During the last two business cycles capacity utilization among mid-sized firms also decreased (−011% and −0.64%) even though quantitatively less than for small-sized firms. Diversely, large-sized firms showed an increase in capacity utilization during the 1996.4–2003.2 business cycle, followed by a decrease (−0.61%) during the most recent business cycle. The latter trend is quantitatively similar to that experienced by the other two size-defined firm groups. The similarity in the decrease in capacity utilization seen during the most recent business cycle is mainly driven by the huge drop (around 20%) in industrial production, observed above all in the first part of 2009. This affected all industries indistinctly.

Lastly, the rebound witnessed in recent quarters is affecting the capacity utilization of firms in ways that seem to favor mid-sized firms in particular (Table 5). However, the exceptionality of the recent recession imposes the need for more careful analysis of up-to-date information on capacity utilization and more data collection to reach some clear-cut conclusions.

Overall, results from the analysis of capacity utilization and industrial production confirm both the consolidation of the Italian industry and the increasing role played by larger firms that feature better human capital, higher investment in R&D, more financial resources and more modern organization; the aforesaid are in the position to challenge the increased international competition more efficiently and better cope with periods of economic crisis.

[16] According to the World Bank 2011 report "Doing Business", Italy ranks 80th (falling 4 positions with respect to the previous year) among 183 countries where the regulatory environment is more conducive to the starting and operation of local firms. More recent data (as of June 2011) show a further worsening of the Italian ranking (http://www.doingbusiness.org/rankings).

9 A Framework to Interpret Trends Among Italian SMEs

The analyses carried out in the previous paragraphs support the hypothesis that Italian manufacturing is undergoing important transformations regarding the adoption of innovations (product, process, organizational), as well as the functional relationships among firms, especially SMEs.

These transformations may represent a non-turning point towards a new production system; the present situation has dominated Italian industry for the last 40 years.

To interpret these transformations, Fig. 8 shows a pyramid the base of which measures the share of small-sized firms in the economy, while along its height technology variables, different stages of the delocalization process (and its functional relationships among firms), and the importance of the underground (or irregular) economy are represented.[17]

At each stage, shown as a section of the pyramid, a different variable configuration is represented. The variables considered are: technology, the types of market with respect to the degree of competition, the role of the underground economy.

The first stage, starting from the base, represents a situation characterized by technology with low levels of innovation, market competition (but with niches of marginal firms), and a large share of the irregular economy.

The second stage includes innovative investments and inter-industry relationships, monopolistic competition, and a smaller share of the irregular economy than in stage one.

Finally, the third stage is representative of an economy characterized by high and increasing levels of innovation, differentiated oligopoly, and a share of irregular economy that mainly reflects dysfunctions in the public sector.

The pyramid-shaped Fig. 8 summarizes the evolution in the role of small-sized firms in Italy and their structure within specific regional contexts.

Going up along the pyramid, technologies become more innovative and inter-industry, while imperfect markets reduce the role of price competition. Further, the irregular economy slowly becomes less important.

As for SMEs, along the pyramid the number of very small-sized firms decreases, while the number of specialized firms increases.

What we are observing during the current economic crisis is a slow but continuous upward movement along the pyramid of Italian industry. This process is not new, but the most recent events represent an opportunity to accelerate the modernization of the Italian economy that cannot be postponed any longer. It should also be noted that the acceptance of the interpretative framework behind the pyramid means that the "upward movement" comes at the cost of a contraction in the production base. In other words, the exit of small-sized firms from the market has long-run implications for the Italian economy. The next cyclical expansion of the

[17] See Favaretto (1995).

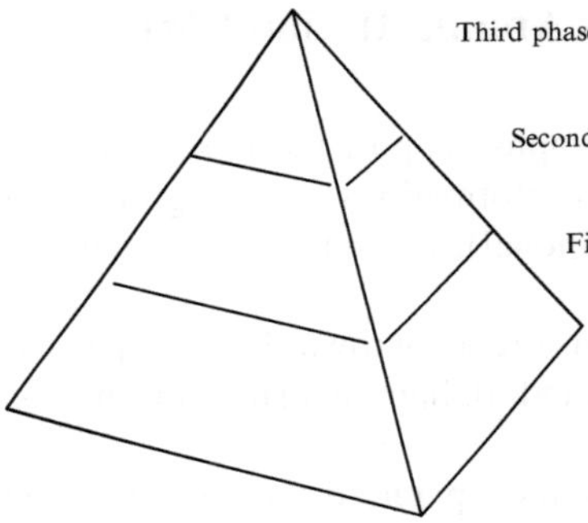

Fig. 8 The evolution of functional relationship among small-sized firms

economy will not likely be enough to create new small-sized firms that will compensate for the ones that disappeared. It will be increasingly necessary to look for new products, services and markets in order to start expanding the Italian production base once more.

10 The Relationship Between Education, R&D and Innovation: Is It a Virtuous Circle or a Bottleneck?

The pyramid framework described in the previous section suggests that there is a strong link between technological innovations and education and training. A larger share of more skilled workers is the condition needed to face the challenge of achieving high-quality production.[18]

In Italy, the supply of specialized workers is traditionally inadequate compared to the needs of firms. This situation, especially within industrial districts, increasingly favors the opening of local labor markets to labor force flows coming from other markets. Further, Italian entrepreneurs often complain about the national school system, considered unable to prepare young generations to meet labor demand influenced by rapidly changing technology. Notwithstanding this, they recognize the importance of well-educated workers, since current and future technologies require a labor force with sounder theoretical knowledge than in past decades. If this were the case, training within firms should be enriched by knowledge that will give them strategic advantages during the adoption of new technologies.

The lack of base knowledge is greater where the economy is unable to remedy the inefficacy of the school system. This mainly occurs in local production systems, characterized by small- and medium-sized firms, where professional training is

[18] For empirical results see Favaretto (2003, 2004, 2006).

strongly conditioned by entrepreneurs focused on few aspects of the production process, quite often limited to technical ones. Therefore, the economic system likely reproduces the context within which former entrepreneurial generations grew up and the development of local economies continues to be based on traditional models of firm organization and market participation.

The current economic crisis is making the match between specialized labor demand and supply more complex. In Italy, both underemployment and human capital underutilization have been increasing; employment has become more precarious among young generations; and the numeric participation of women in the labor market has worsened, as have job opportunities for university graduates and researchers.

However, innovation is also affecting local systems in a positive manner. The upward movement along the pyramid of Fig. 8 implies a delocalization process based on complex functional relationships among firms driven by an accelerating adoption of innovations. This process requires the presence of coordination and guarantees for those production phases characterized by higher quality content. More often small-sized firms, by guaranteeing the quality of their production phase, efficiently provide coordination for the collaboration among firms.

11 Conclusion

This Chapter analyzed the medium- and long-run state of Italian SME dynamics and focused on new phenomena that occurred during the last decade. It described how the labor division among firms during the last 30 years influenced the numeric growth of Italian businesses, while leaving the technology content of their products almost unaffected, i.e. the majority remained concentrated in traditional industries characterized by intermediate technologies.

Higher technological, organizational, and product-quality levels characterized production delocalization, even though not all industries have experienced the same intensity of innovation. Notwithstanding this fact, the selection process among industries has been severe, but the Italian industrial districts have survived the change in the global competitive context and SMEs have increased their participation in economic activities characterized by advanced technologies. Indeed, the current crisis is penalizing those firms that did not restructure and rationalize, or invest in R&D. Further, it favors the consolidation of the activities of firms that were previously outsourced.

Small-sized, but young and well-organized firms seem to be in a position to better take advantage of the opportunities provided by the economic crisis and the redefinition of their competitive position and market share. Firm consolidation takes the form of an increase in firms' average size, in the number of limited liability companies, and in higher levels of capacity utilization.

These trends imply new functional relationships among firms that may drive the Italian economy towards an industrial structure different from the one that prevailed until recently. In the new economic context small-sized firms will be asked to take on complex organizational responsibilities, while small-sized firms located at the base of the functional-technology pyramid will progressively disappear. The latter, however, does not necessarily mean a decrease in the total number of small-sized businesses, if they are able to take advantage of the opportunities provided by the development of new complex functional relationships among firms.

Therefore, the Italian production system needs a higher level of innovation, especially among small-sized firms and the economic crisis could provide the right opportunities to modernize the economy. Indeed, according to the pyramid-shape framework, an increase in the intensity of innovation among small-sized firms will result in a more ample diffusion of the consolidation process, that will also change their role within the new and more complex industrial organization. The delocalization of production will be characterized by new functional relationships among firms, and bigger and more diversified production chains. Industrial district externalities will become less important, but will not be canceled by the increasing role of inter-industry externalities.

References

Accetturo A, Giunta A, and Rossi S (2011) Le imprese italiane tra crisi e nuova globalizzazione, Questioni di Economia e Finanza, Occasional Paper no. 86, Banca d'Italia

Banca d'Italia (2010) Economie regionali. L'economia delle regioni italiane, no. 85, Luglio

Becattini G (1989) Il distretto industriale marshalliano come concetto socio-economico, Stato e Mercato, no. 25

Brandolini A, Bugamelli M, Torrini R (2010) Ristrutturazione delle imprese e produttività, In: Istat – X Conferenza nazionale di Statistica, Roma

Calcagnini G, Favaretto I (2011) The economics of small business. An international perspective. Physica Verlag, Heidelberg

European Commission (2010) European SMEs under pressure, Annual report on EU small and medium-sized enterprises 2009, Brussels

Fadda S (2010) Sviluppo locale e obiettivi della programmazione, Argomenti, no. 29, maggio-agosto

Favaretto I (1995) Mercati imperfetti e decentramento produttivo. ASPI/INS-EDIT, Urbino

Favaretto I (2003) Factors of regional competitiveness, In: International conference of regional studies association (Reinventing Regions in Global Economy), Pisa 12–15 Aprile, atti della conferenza

Favaretto I (2004) Il mercato del lavoro: dinamiche, tendenze, specificità. In: Diamanti I, Ceccarini L (a cura di), Mappe e scenari della società regionale, Primo rapporto, Consiglio Regionale delle Marche – La Polis

Favaretto I (2004b) Innovazione e ambiente, fattori di competitività delle economie regionali. Argomenti 8:19–45

Favaretto I (2006) Distretti e internazionalizzazione. Il caso delle Marche, Sinergie, no. 69

ISTAT (2010) La demografia d'impresa. Anni 2003–2008, Statistiche in breve, settembre, www.istat.it

ISTAT(2011) http://www.coeweb.istat.it/

Schumacher EF (1973) Small is beautiful: economics as if people mattered. Harper and Row, New York
TrendER (2009) Osservatorio congiunturale sulla micro e piccola impresa in Emilia Romagna, http://www.cnaemiliaromagna.it/trender.php
World Bank (2011) Doing business 2011: making a difference for entrepreneurs. World Bank, Washington, DC

Product Innovation and Corporate Governance in Turbulent Times: Evidence from Italian SMEs

Marco Cucculelli

1 Introduction

Decisions of strategic relevance are expected to be based on long-term planning horizon and to vary only marginally with short term exigencies (Devinney 1990). Yet, conventional wisdom argues that short-term factors – especially negative ones such as low earning, poor economic conditions or short-term changes in GDP growth – do affect the timing of these decisions, either by speeding them up or slowing them down.

This effect may vary considerably among firms, either because companies adopt different strategic responses to the same shock, or because many decisions are interweaved and insist on the same time frame. For example, when economic conditions deteriorate, family successions can be postponed, as the founder wants to leave the company in good shape (Adams et al. 2009), or speeded up in order to provide a different way-out of the downturn. Similarly, product development can be halted in a downturn, or speeded up to help the firm to be positioned when markets recover (Kauffman 2009).

This paper analyses the impact of short term factors on two decisions of strategic relevance for the firm: (1) the new product introduction and (2) the founder/heir succession. Specifically, the paper examines if and to what extent the business cycle affects the firm's decision process concerning the introduction of a product innovation or a change in the firm governance structure.

The paper uses longitudinal data on a sample of 142 Italian family firms. Data on the number of products per firm, the year of introduction of each product and the

M. Cucculelli (✉)
Department of Economics and Social Sciences, Università Politecnica delle Marche, Piazzale Martelli, 8, 60100 Ancona, Italy
e-mail: m.cucculelli@univpm.it

G. Calcagnini and I. Favaretto (eds.), *Small Businesses in the Aftermath of the Crisis*, Contributions to Economics, DOI 10.1007/978-3-7908-2852-8_6,

year of family succession are merged with other firm-specific information and financial data, in order to describe the evolution of the firm's decision structure over time. Factors affecting the time it takes to adopt a new product and to make a succession are examined using a conditional risk set model. Differently from the Cox hazard model, this duration model allows to take into account the multiple ordered occurrences of similar failure, i.e. additional failures of the same type that cannot occur before the first. It therefore allows to stratify data by risk level (each firm can adopt more than a product and have more than a succession event in its life) and to cluster observation by firms.

Preliminary evidence shows that economic downturns do affect firm decisions substantially, as both product introduction and succession are delayed after the insurgence of the recession. The probability for both the events to occur increases steadily as the negative effects of downturns tend to vanish, but at different rates: delays in succession prove to be shorter than those in product adoption. Furthermore, it is more likely to observe that succession causes new product introduction than vice-versa: the probability to introduce a new product after a succession is higher than the probability for the founder to step down after the conclusion of the product introduction process.

2 Literature and Hypothesis Development

In recent times, there has been extensive discussion of the notion that "recessions are times when productivity improving activities are undertaken because of their temporarily low opportunity costs" (Caballero and Hammour 1994) or because the increase on the probability of bankruptcy due to bad times raises the pressure of managers to initiate organisational changes (Nickell et al. 2001). In these periods, the likelihood that a firm will be forced out of business is large, challenging employment stability (Nickell et al. 2001), accelerating the job turnover (Davis and Haltiwanger 1992) and firm turnover (Oulton 1987), or firm survival (Nishimura et al. 2005; Alvarez and Görg 2009).

2.1 Product Innovation

Many products that we know as stunning successes were actually launched during economic downturns. Many of them found success precisely because, not just in spite of, unusual economic conditions, as companies can see market opportunities where others fail to look (Kauffman 2009). Even new companies – not just products – are likely to start in turbulent times: more than half of the companies on the 2009 Fortune 500 list were launched during a recession or bear market, along with nearly half of the firms on the 2008 Inc. list of America's fastest-growing companies (Kauffman 2009).

The recent financial crisis has, among other things, reduced the short term willingness of companies to invest in innovation. However, the strong decline in innovation propensity has been counterbalanced by a greater occurrence of major innovations as a result of the change in underlying economic paradigm (Kleinknecht 1990). Incremental as opposed to major innovations are more frequent during periods of steady economic growth (Freeman et al. 1982).

Several justifications motivate the focus on the appearance of a new product as the output of innovation activity. It has been observed that the production of new goods reflects innovation activity which has been successful (Hall et al. 2009; Geroski and Mazzucato 2002). On the opposite, the adoption of traditional indicators, such as various R&D indexes, does not permit to assess the effectiveness in the use of these resources (Flor and Oltra 2004). Focusing on R&D has the drawback of overlooking all innovations arising via informal mechanisms (i.e., learning by doing, interactions with partners, etc.) instead that via organized labs or structured research activity (Flor and Oltra 2004). Similarly, also the use of patent counting as a proxy for innovation is not exempt from criticism, as not all patents have 'commercial' implementation either not any discovery is patented (Arundel and Kabla 1998; Becheikh et al. 2006). Finally, it has been stressed that innovation indicators based on innovation output are easier to collect and can offer disaggregated information content (Kleinknecht et al. 2002; Corsino 2008). Apart reasons linked to the pro and contra of alternative innovation indicators, we choose to rely on new product introduction to measure innovation activity since it signals a modification into the existing product mix of a firm: this aspect best suits to a dif fuse feature of today manufacturing industry, that is the multi-product nature a of the firm.

The existing empirical literature does not provide unequivocal evidence concerning the probability of product adoption according to short term economic conditions (Devinney 1990; Alvarez et al. 2006; Bhattacharjee et al. 2007). As a general orientation, product innovation reflects the accumulation of innovative capabilities within the firm; therefore, the introduction of a new product could be modeled as a random occurrence following the randomness of inventions and scientific discovery which occur in the firm. However, product introduction can also be seen as a defensive mechanism to face the crisis: this makes product innovation pro-cyclical (Devinney 1990). If this is true, the propensity of introducing a new product should be higher following recessions. It has to be taken into consideration that the introduction of a new product requires some time to be adopted. If, for example, a product takes 1 year to be ready for the market, the manager will make the decision of innovating in the year of the crisis while the reaction itself will only be recorded in the following year. Following this approach the hypothesis can be expanded including also the effect of the cycle during the years before the crisis. Summing up, a product-introduction hypothesis can be tested by assessing if (1) there is an effect of the economic cycle in the new product introduction and (2) the number of new product introductions is higher in negative economic periods and in the following years than in periods of expansion.

2.2 Family Succession

A second hypothesis explores the possibility that the successions are influenced by the economic cycle. It is not so rare that, if the economy is in a difficult period and the firm performance gets worse, the founder will prefer to postpone the succession to an heir until the firm reaches a better chance of surviving. This effect should be amplified in family successions, as the founder values control over the succession and wants to leave the firm in good conditions (Adams et al. 2009; Anderson and Reeb 2003; Wasserman 2003). Therefore, we expect that if the economy is in recession, there will be very few successions while, in the years after the economic crisis, the number of successions should rise.

The suggestion that some of a firm's characteristics (its current performance and future perspective) might affect the timing of the transfer and the type of CEO is not new in the literature on the transfer of assets and business. Focusing on the optimal timing of succession, Kimhi (1994) and Miljkovic (1999) show that the optimal transfer time varies systematically with family and firm characteristics, as well as with other variables defining the critical level of the founder's utility from the firm transfer to a succeeding child. Similarly, the hypothesis that an optimal selection of the CEO may emerge endogenously from the founder's characteristics and the firm's performance has recently received significant empirical support. Adams et al. (2009) show that a founder may deliberately choose to delay the transfer until the firm's performance has reached a level high enough for the heir to have a better chance of improving its future operations. They explore this issue by assessing the extent to which the firm's past performance, whether bad or good, helps in predicting a future change of command in which a founder steps down. They find evidence that a past good performance increases the likelihood that a founder will leave the firm in the near future, and that this result is consistent with two potential factors: (1) founder–CEOs may value control over their succession more than nonfounders, and (2) founder–CEOs may want to leave their companies 'in good shape'. They conclude that this result has a direct implication for the evaluation of a firm's performance after the founder's exit: if performance is mean-reverting and the founders leave at its peak, performance is likely to decline after the succession even when the inherited control is not bad for performance. In addition to the above "controlled-succession" effect, Wasserman's (2003) "paradox of entrepreneurial success" shows that a founder's success in achieving critical milestones, such as the successful completion of product development, makes it more likely that he will step down. Both these stories predict that founder–CEOs will retain the status until the negative performance has disappeared. And the finding that founders are more likely to leave when their firms are doing well suggests that founder–CEO succession can be very different than the succession of professional CEOs, in the sense that founder succession to an heir can be delayed, whereas external CEO succession to another professional CEO can not.

3 The Empirical Model

Multiple failure-time data or multivariate survival data arise from time-to-occurrence studies when either two or more events (failures) occur for the same subject, or from identical events occurring to related subjects such as family members (Cleves et. al. 2004). In these studies, failure times are correlated within cluster (subject or group), violating the independence of failure times assumption required in traditional survival analysis.

Failure events should be classified according to (1) whether they have a natural order, and (2) whether they are recurrences of the same types of events (Cleves, 1999). The simplest way of analyzing multiple failure data is to examine time to first event, ignoring additional failures. This approach, however, is usually not adequate because it wastes possibly relevant information. Alternative methods have been developed that make use of all available data while accounting for the lack of independence of the failure times. Two approaches to modeling these data have gained popularity over the last few years. In the first approach, the frailty model method, the association between failure times is explicitly modeled as a random-effect term, called the frailty, where frailties are unobserved effects shared by all members of the cluster. In the second approach, the dependencies between failure times are not included in the models. Instead, the covariance matrix of the estimators is adjusted to account for the additional correlation. These models, which are called "variance-corrected" models, are easily estimated by major statistical packages.

There are several approaches to the analysis of ordered events. The principal difference between these methods is in the way that the risk sets are defined at each failure time. The simplest method to implement them follows the counting process approach. The basic assumption is that all failure types are equal or indistinguishable. The problem then reduces to the analysis of time to first event, time to second event, and so on. Thus, the risk set at time t for event k is all subjects under observation at time t. A major limitation of this approach is that it does not allow more than one event to occur at a given time. A second model is based on the idea of marginal risk sets. For this analysis, the data is treated like a competing risk dataset, as if the failure events were unordered, so each event has its own stratum and each individual appears in all strata. The marginal risk set at time t for event k is made up of all subjects under observation at time t that have not had event k. A third method the is known as the conditional risk set model. The data are set up as for the counting processes method, except that the analysis is stratified by failure order. The assumption made is that a subject is not at risk of a second event until the first event has occurred and so on. Thus, the conditional risk set at time t for event k is made up of all subjects under observation at time t that have had event $k - 1$. There are two variations to this approach. In the first variation, time to each event is measured from entry time, and in the second variation, time to each event is measured from the previous event.

The model we have chosen for the estimates is the first variant of the conditional risk set model (time from entry) since the failure events are ordered: the firm is not at risk of a further product introduction or founder succession until the previous one (succession or product introduction) has been verify. Differently from the Cox hazard model, this duration model allows to stratify data by risk level (each firm can adopt more than a product and have more than a succession in its entire life) and to cluster observation by firms, in order to group firm-level data in the same cluster. A survival analysis method is preferred than other linear models since, also in the case of successions, it is hard to think that the hypothesis of normal time distribution subordinated to our covariates can be held (Cleves 1999).

3.1 The Datasets

The Fondazione Merloni dataset provides financial information on 142 small and medium-sized companies. Financial data has been merged with survey information obtained from a questionnaire aimed at collecting longitudinal data about product introductions and successions. Finally, survey data and financial data have been merged with the index of industrial production for the Italian manufacturing industry obtained from Istat – the Italian Central Institute of Statistics.

After a preparatory study, based on the literature and an interview outline that was pretested in a subsample of 16 companies (not included in the final study), questionnaire-based interviews had been conducted by a group of six research assistants from Merloni Foundation in a 2-h individual meeting with selected interviewees in all companies. The questionnaire was addressed to the "person in charge of major company decisions," i.e., the Chief Executive Officer (CEO), the chairman/president, or highest ranking executives, who were assured of the confidentiality of their responses. These individuals were targeted because they are the most knowledgeable persons in their firms. Analysis of the formal titles of responding executives showed that 78% were owners or CEO/president, 16 were general managers (family or external), and 6% other senior executives.

Interviewed companies compete in 21 three-digit NACE sectors. As the sample composition in terms of sectoral breakdown closely resembles the structure of the Italian manufacturing industry, we are confident that results may reflect the experience of a broader firm population (i.e., medium-sized firms that are the core of the Italian manufacturing industry). The sample modal class is between 50 and 100 employees and the entire size distribution is symmetric with respect to the modal class. Average firm size is about 125 employees and firms are 34 years old on average. These data are very close to average values computed for medium-sized companies in the manufacturing industry by the Italian Institute of Statistics (Istat 2009).

The type of new product introduction that we have considered involves a radical change in the product portfolio of the company and a substantial enhancement of the firm capabilities (technical and commercial) for the realization of the product. We followed the suggestion put forward by Bernard et al. (2010), by considering a new product a different five-digit NACE category in the firm product portfolio. We do not consider small refinements or negligible enhancements in the existing features of the product. Data on product introduction has been obtained by asking the interviewed persons the following questions. "After having listed all the products present in product portfolio, please give separately for each product: (1) a detailed description of the product characteristics and a comparison with other products in portfolio; (2) the year of introduction of the product." By using the detailed product description from question (1), we coded manually each product in a five-digit classification in order to define the whole firm product portfolio. At the date of the survey, half the firms in the sample had only one product in its current portfolio. The remaining groups had two (21% firms), three (26%), and four or more products (4%) in the product portfolio. In spite of the relevance in terms of number of firms (49%), single-product companies' account for a minor share of whole sample sales (19.6%), whereas multiproduct companies are responsible for about 70% products and 80% total sales.

With respect to the family succession event, the questions asked were about (1) the type of current management (heirs, an acquiring company, other external managers) and (2) the date when the succession took place. Therefore, we asked about the changes in the management, regardless of the status of company ownership and control. Two main distinctive features of Italian family businesses support this assumption: an almost complete overlap exists between ownership and management, and management changes normally trigger subsequent changes in company control (from the founder to heirs), rather than the reverse (Cucculelli and Micucci 2008).

3.2 Results

Economic downturns have been identified by changes in the industrial production index. Empirical estimates has been carried out using three different variables indicating "recession": (1) a dummy that is set equal to one if the yearly change of the index is negative and zero otherwise; (2) a dummy that is set equal to one if the yearly change of the index is lower than 0.5% and zero otherwise; (3) the integer value of the yearly change rate of the index. The reason for the less restrictive value of 0.5% under (2) allows for a slightly more equilibrated sample of negative and positive years, as managers may consider a bad economic period not only when the production index has a negative value, but also when it simply declines. In the empirical estimates, we have used the first variable, i.e. the dummy variable approach for the yearly change of the index. Both the second and the third variables have been used as a robustness check, not reported for the sake of brevity.

Recessions do affect product introduction when they happen: the hazard ratio in column 1 shows that the introduction of a new product is 71% less likely in recession times than in normal ones (the coefficient of the first variable is 0.29). However, in the years following recession, the introduction of new products is between 55% and

Table 1 Conditional risk set model from time to entry. Failures: new product introduction and succession. Δprod = change of the industrial production index

	Failure = new product	Failure = succession
Δprod in t0	0.29***	0.14**
Δprod in t − 1	1.55**	2.11**
Δprod in t − 2	1.69*	1.13
Δprod in t − 3	0.98	1.44**
Number of products	4.88***	1.09*
Type of succession[a]	1.12	2.24*
Years from last succession	0.84	0.67*
Years from last product introduction	1.55*	1.01
Young workforce	2.37***	0.73*
Succession t0	0.69***	
Succession t − 1	0.80*	
New product introduction t0		0.19**
New product t − 1		0.27**
Log-pseudolikelihood	−128.75	−149.72
Wald chi2 test	243.71**	204.43***

Legend: Δprod = Percentage change in the industrial production index. Δprod = 1 when the change in the index is negative and 0 otherwise

[a]Type of succession = 0 if the firm has never had a succession; type = 1 if the firm has had a family succession; type =2 if the firm has hired an external CEO; type = 3 if the firm has been sold

All the models are run using the option cluster(id) to obtain a robust estimation for addressing multiple records and the option strata(str) to stratify for the risk level. All the models use the variable "group" (not reported) for control. Statistical significance levels are: 0.01 '***', 0.05 '**', 0.1 '*' and 0.15 '.'

70% more likely than average times, thus confirming the pro-ciclicality of product introduction. So, downturns are likely to affect the propensity of the firm to innovate, but with a lag due to the time it takes to the firm to react to the negative economic period. This makes the product introduction to lead the cycle (Table 1).

Similar evidence is observed for successions, whose occurrence is sharply reduced (even more than product introduction) by a recession, whereas its likelihood increases substantially in the second period.

With respect to the covariates, both a larger number of products in portfolio and a higher share of young workforce increase the likelihood to introduce a new product: this suggests increasing returns in the product portfolio management and a high propensity to introduce new product among young firms. This evidence is less robust in the case of successions.

The occurrence of a succession has a negative effect on product introduction. The hazard rate concerning the introduction of a new product shows that the likelihood of this event to occur decreases by 31% after a succession (column 1, coefficient = 0.69). Therefore, when a succession happens, the probability of observing the introduction of a new product decreases. The likelihood to introduce a new product slightly increases in the second years after the succession, but it still remains lower that the average (the hazard ratio of the variable is 0.80).

Similarly, the introduction of a new product does affect the probability of a succession substantially; and this effect is far more significant than in the case of a succession affecting new product introduction. Each year characterized by a product introduction is hardly never followed by a succession. Therefore, it is more likely to observe a new product introduction after a succession that vice-versa.

4 Conclusions

The empirical analysis shows that economic downturns affect firm's decisions substantially, as they change the timing of product introduction and succession or stop their adoption at all. The probability for both events to occur increases steadily as the negative effects of downturns vanish, but at different rates: delays in succession are shorter than those in product adoption. Finally, succession induces new product introduction only when the first happens in favourable years, whereas new product introductions always reduce the probability to have a succession. However, this negative impact is less evident in positive years, when the pro-cyclical nature of product introduction is more evident.

References

Adams R, Almeida H, Ferreira D (2009) Understanding the relationship between founder-CEOs and firm performance. J Empirical Finance 16:136–150

Álvarez R, Benavente JM, Crespi G (2006) Does economic crisis spur organizational change? Evidence from Chilean manufacturing plants, available at: http://ssrn.com/abstract=940400

Alvarez R, Görg H (2009) Multinationals and plant exit: evidence from Chile. Int Rev Econ Finance 18(1):45–51, January

Anderson R, Reeb D (2003) Founding-family ownership and firm performance: evidence from the S&P 500. J Finance 58:1301–1328

Arundel, A., Kabla, I. (1998) What percentage of innovations are patented? Empirical estimates for European firms, Research Policy 27:127–141

Becheikh N, Landry R, Armara N (2006) Lessons from innovation empirical studies in manufacturing sector: a systematic review of the literature from 1993–2003. Technovation 26:644–664

Bernard A, Redding S, Schoot P (2010) Multi-products firms and products switching. Am Econ Rev 100(2):444–448

Bhattacharjee A, Higson C, Holly S, Kattuman P (2007) Macroeconomic conditions and business exit: determinants of failures and acquisitions of UK firms. Center for dynamic macroeconomic analysis working paper series, available at: http://ssrn.com/abstract==1011737

Caballero R, Hammour M (1994) The cleansing effect of recessions. Am Econ Rev 84 (5):1350–1368

Cleves MA (1999) Analysis of multiple failure-time survival data. http://www.stata.com/support/faqs/stat/stmfail.html

Cleves MA, Gould W, Gutierrez RG (2004) An introduction to survival analysis using Stata. Stata Press, College Station

Corsino M (2008) Product innovation and growth: the case of integrated circuits. LEM working paper 2008–02, Sant'Anna School of Advanced Studies, Pisa

Cucculelli M, Micucci G (2008) Family succession and firm performance: evidence from Italian family firms. J Corp Finance 14(1):17–31

Davis S, Haltiwanger JC (1992) Gross job creation, gross job destruction, and employment reallocation. Quar J Econ 107(3):819–863, August

Devinney TM (1990) New products over the business cycle. J Prod Innovat Manag 7:261–273

Flor ML, Oltra MJ (2004) Identification of innovating firms through technological innovation indicators: an application to the Spanish ceramic tile industry. Res Policy 33:323–336

Freeman C. Soete L. Townsend J. (1982) Fluctuations in the Numbers of Product and Process Innovations 1920–1980, Paper for the OECD Workshop on Patent and Innovation Statistics, June 28–30, Paris

Geroski P, Mazzucato M (2002) Learning and the sources of corporate growth. Ind Corp Change 11:623–644

Hall BH, Lotti F, Mairesse J (2009) Innovation and productivity in SMEs: empirical evidence for Italy. Small Bus Econ 33:13–33

ISTAT (2009) Annual Report, Roma

Kauffman Foundation (2009) The economic future just happened. Kauffman 2009 Fast Facts

Kimhi A (1994) Optimal timing of farm transferal from parent to child. Am J Agr Econ 76:228–236

Kleinknecht A. 1990 "Are there Schumpeterian Waves of Innovations?", Cambridge Journal of Economics, vol. 14, n. 1, pp. 81–92.

Kleinknecht A, Van Montfort K, Brouwer E (2002) The non-trivial choice between innovation indicators. Econ Innov New Technol 11(2):109–121

Miljkovic D (1999) Optimal timing in the problem of family farm transfer from parent to child: an option value approach. J Dev Econ 61:543–552

Nickell S, Nicolitsas D, Patterson M (2001) Does doing badly encourage management innovation? Oxford Bulletin of Economics and Statistics, Department of Economics, University of Oxford, 63(1):5–28

Nishimura K, Nakajima T, Kiyota K (2005) Does the natural selection mechanism still work in severe recessions? Examination of the Japanese economy in the 1990s. J Econ Behav Organ 58(1):53–78

Oulton N. (1987) Plant Closure ad the Productivity Miracle in Manufacturing, National Institute Economic Review, 53–59, August

Wasserman N (2003) Founder-CEO succession and the paradox of entrepreneurial success. Organ Sci 14:149–172

The Internationalization of Italian SME After the Crisis: New Opportunities or New Threats?

Rosalba Rombaldoni

1 Introduction

International economic relationships are going through big changes, due to fragmented economic activity and the integration of commercial flows, with strong effects on the world economy and international competition.

The recent economic and financial crisis has violently shaken markets and relations among the main players on the economic scene, such as firms, consumers and institutions. If up until now survival was the main concern, today new paths of sustainable growth have to be found in the long term. According to many authors the right direction is that of internationalization, especially in the light of wide empirical evidence suggesting a significant and positive relationship between international trade, income and national growth.

However, the facets of such a process are numerous, complex and often closely related: delocalization, fragmentation of the productive process, technology and innovation transfers and globalization of production chains, are just some of the aspects that point to internationalization as a growth strategy.

Production delocalization is one of the most frequent answers to the growing pressures of globalization and translates into a complex process of organization that redefines relations between firms, altering their geography and profit opportunities, according to changes in demand and technological advances. Chain value are increasingly globally widespread and through them relationships between businesses develop, so leading to international integration of production.

The link between internationalization and innovation has been investigated by many authors who demonstrate how there is more innovation among international

R. Rombaldoni (✉)
Dipartimento di Economia, Società, Politica, Università di Urbino "Carlo Bo", via Saffi 42, 61029 Urbino, Italy
e-mail: rosalba.rombaldoni@uniurb.it

G. Calcagnini and I. Favaretto (eds.), *Small Businesses in the Aftermath of the Crisis*, Contributions to Economics, DOI 10.1007/978-3-7908-2852-8_7,

firms than in those operating in the domestic market.[1] The positive effect on commerce is expressed by the capacity to innovate and the many incentives to do so, but technological transfers and the creation of economies of scale have the strongest impact on innovation. International firms generate more knowledge than national ones and this is due on the one hand to massive knowledge input, and to greater access to external ideas on the other.[2]

A relevant aspect to be observed in international markets is the spontaneous aggregation among firms into districts or a global production chain. In the first case, this is a profound evolutionary transformation: the informal links of sub-contracting and know how that keep together a multitude of small firms and a few large enterprises in the same productive tissue, during the economic crisis have been under strong pressure and have subsequently collapsed. This has had serious consequences on local work systems, particularly for highly specialized productions, with stiff international competition regarding the cost of labour and availability of raw materials. Big changes on the international scene have noticeably modified the nature of subcontracting, with the smallest and more fragile firms being pushed out. The context that has gained ground is that of global chains, where production by multinational companies is organized on a wide geographical base, without local connections to the area.

Besides the advantages of directly participating in internationalization, the global chain has an important impact on gaining new knowledge, in exchanging know-how, productive specialization, and joint trials. It also allows firms to grow according to different patterns, from strategic and quality upgrading thanks to improvements in production efficiency, to a widening of the range and quality of products.[3]

The aim of this work is to identify the behaviour of Italian SMEs in an international context in order to outline paths of sustainable development. The above aspects will be analyzed, with the aid of recent theoretical literature and significant empirical data.

2 Fragmentation of the Productive Process and Globalization

International fragmentation of production, as defined by Jones and Kierskowski (2000) and Venables (1999), can translate into processes of vertical and horizontal integration. In the first case the productive structure of the mother firm is reproduced in other geographical locations through a FDI (direct foreign investment) in order to gain better access to local markets, and this is typical of developed

[1] Onodera (2008).

[2] Pittiglio et al. (2009).

[3] Unicredit, “Rapporto Unicredit sulle piccole imprese”, VII edition 2010–2011.

countries; in the second case the productive process is fragmented and relocated elsewhere through a FDI or lighter modes of internationalization (such as delocalization, sub-contracting, joint ventures) with firms belonging to low cost areas. Less developed and developing countries are usually involved.

However, strong evidence of increasing of task trade is hard to come by. First of all data is collected and reported as gross flows, instead of foreign value added, making it difficult to identify where tasks were effectively performed. Moreover, some of this trade, especially when it involves businesses services, does not leave any paperwork. Lastly, we cannot forget the complexity of the adopted mode (FDI, joint ventures, subcontracting). In the course of the work we will try to understand the impact of similar factors on Italian foreign trade, and to what extent firms of different sectors are involved.

2.1 The Change of Comparative Advantages for Italy

Principal theories of international trade have given different explanations about countries' specialization and evolution of exchange flows, from factors endowment to economies of scale, from the size of demand to the role of historical accident in the presence of external economies, to endogenous factors that reflect the localization choices of firms. However, whatever the cause, similar comparative advantages reflect on commercial and sectorial trade balances (Bugamelli 2001), and in the work by Foresti, Trenti (2006)[4] it is clear that Italy has taken a new position in international labour division: in the period from 1996 to 2004, comparative disadvantage is high for primary goods, we lose ground for consumption goods, a better performance is given by investment goods while signs of improvement are present for intermediate goods (whose exports go from 44.2% in 1996 to 46.1% in 2004, vs. imports equal to 45.8% and 43.3%), both for semi-manufactured goods and parts and components. The size of the latter also goes down where imports are concerned, in parallel with an increase in imports of consumption goods.

Therefore Italy goes from being a specialized country where it comes to finished products, to a country that places itself at the beginning of the production chain where intermediate goods are concerned (used by other countries), keeping at the same time a comparative advantage in the production of investment goods.

The same trend was found in a geographic area such as the European Union: for the same period Italy has a comparative advantage for parts and components for investment goods and decreases the disadvantage for semi-finished goods. This

[4] This is the work of Tattara, Corò, Volpe (2006). The described data are the contribution to the balance and composition of Italian trade for primary, intermediate (semi-finished parts and components), final goods (investment and consumption). The indicator is that of Lafay, which measures the contribution of a sector to the total trade balance, through normalizing the weight of the different sections. For a detailed description, see the cited work.

offsets the considerable reduction in the contribution of consumption goods to the balance; for intra-ue15 trade Italy holds anyway a comparative disadvantage.

For areas where delocalization is strong there are remarkable differences: in close countries not belonging to the EU (eastern Europe, Turkey and the Mediterranean), for the same time span, we notice a growth in comparative advantages for intermediate and investment goods and a reduction of advantages for consumption goods. The area including the rest of the world denotes a generalized and marked worsening for products at the end of the chain. This coincides with many national and international studies (Gereffi 1994; Crestanello and Dalla Libera 2003), according to which *buyer-driven* or commercial chains prevail in the case of Asia, whereas in Europe chains are based on internationalization of subcontracting. In the first case the local producer gets the final products without receiving the inputs to be assembled, in the second case the customer firm supplies semi-finished products and components which are assembled in low labour cost countries.

Useful information can be drawn from the same kind of analysis, at sector level[5]: within manufactured goods, the main factor affecting balance is instrumental mechanics, which is the major comparative advantage of the country. Specialization in some sectors has given rise over time to positive externalities that promoted "specialized supply sectors," in other words production of machines and equipment that are functional to the sector down the chain (De Nardis 1997).

Over time this has been equivalent to a cross fertilization process, with continuous interactions between user and producer, phenomena of learning through use and significant innovations. Comparative advantages of this type of production have exceeded in the considered time span those of other commodities, typical of *made in Italy*, such as the aggregate of sectors within the fashion system, textile-clothing and shoes. The notable fall of performance of intermediate goods is surely ascribable to products down the chain, in direct contact with the customer (clothing, footwear and also textiles) while basically no change is registered for intermediate goods.

Better performance of intermediate goods is evident also for sectors that give a negative contribution to the balance, such as motor vehicles and electronics, while food and pharmaceuticals went from a disadvantageous situation in 1994 to an advantageous one in 2006. The interpretation of such changes can be twofold: the positive contribution to the balance in the phases up the chain is referable to the period of crisis of national industry that decreases imports of intermediate goods and components, in order to find new markets. Secondly, the better performance of manufacturing up the chain leads again to processes of delocalization (more or less strong) of firms that keep stoking their production with semi-finished goods and components produced in Italy. For the food and the fashion sector, both interpretations can be considered valid.

The picture so far outlines the background for Italian firms, highlighting its strengths (comparative advantage and disadvantage) in terms of international presence from the geographical and sectorial points of view. Changes in their

---

[5] It is again the study by Foresti, Trenti in Tattara, Corò, Volpe (2006).

competitive positions depend jointly on participation in internationalization processes and fragmentation of production.

3 The Possible Modes of Internationalization

The term internationalization can assume different meanings, and in a first classification we distinguish between the commercial (import–export activity) and the productive one (FDI). However between the two extremes we have a whole series of intermediate activities that go from commercial or technical agreements to franchising, from relocalization and delocalization to contracting and subcontracting, from joint venture projects, outsourcing, to participation in global chains and the birth of foreign enterprises thanks to the initiative of Italian entrepreneurs with local capital.[6]

The data from the EU Report on SME internationalization in 2010[7] show how Italy is almost in line with the European average as far as exports are concerned, but well below when FDI, technical cooperation and foreign subcontractor relationships are considered, even ending up last (Fig. 1).

Besides any considerations beyond this evidence of the relative international weakness of Italy, the logical next step is to thoroughly investigate the various modes of internationalization that takes different forms and dimensions according to the choices of the individual entrepreneur and the distinctive characteristics of the productive system.

3.1 *Market Shares, Qualitative and Geographic Positioning of Italian Exports*

Exchanges of goods and services represent the most common form of interaction with foreign countries, and Italy is among the first ten countries of the world both for export and import.[8]

[6] Some terms need a more specific definition: *outsourcing* is related to the production of a company that operates outside national borders, whereas *delocalization* (often used as a synonym of *outsourcing*) is when part of the productive process leaves the country but not the company because it is carried out by a foreign subsidiary company or a company that is part of the same group. *Joint venture* instead is a more structured collaboration with foreign enterprises.

[7] Modes of internationalization within the Report are: export, import, FDI, international agreements of technical cooperation and subcontracting relationships with foreign countries. Enterprises in the survey are 9480, in 33 European countries (27 EU members and 6 non members: Croatia, Iceland, Liechtenstein, Former Yugoslav Republic of Macedonia, Norway and Turkey). Firm sizes are micro (1–9 employees), small (10–49), medium, (50–249) and there are 26 sectors of activity.

[8] Unicredit Report 2010: Italy places itself at the seventh place in the world for exports and at eighth for imports.

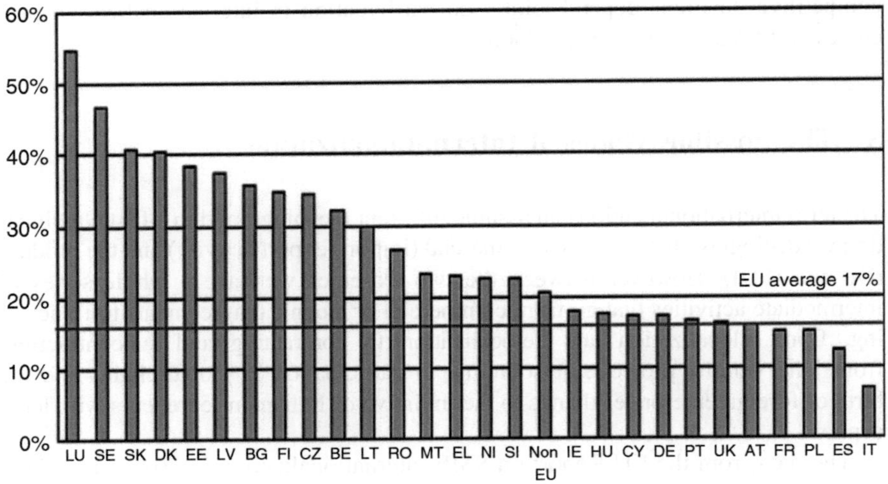

Fig. 1 Presence of SME (in %) with FDI, technical cooperation a/o foreign subcontractor by country, 2006–2008 (Source: Internationalization of European SMEs, final report, European Commission Enterprise and Industry (2010))

Figure 2 gives an idea of the evolution of competitiveness of the market share for Italian exports. The value of the Italian share worldwide goes from 5% to 3%, in the 20 years from 1990 to 2010, whereas in the EU-15 drops from 11.3% to 10%.[9] This loss is due to the rise of emerging countries and competitive pressure from China, together with aspects that deal with our specialization model, oriented towards products whose world demand has been growing below average. The negative aspect of the commodity composition of our export translates into a low income elasticity of demand, combined with a weak presence of product innovation that might modify consumption patterns of families.

Export unit values (as a ratio between value and quantity of exported goods) have fallen less with regard to prices[10] and this trend would confirm the improved quality of the exported product. Over the last 10 years, the increase of export unit values has been notable, even compared with the EU average, and there are two possible explanations.

[9] The majority of studies on Italian export (this included) do not consider the last revision of Istat on export volume data: the old series were deflated with unit values, whereas now export prices are used. The new data show a strong increase (and not a decline) of Italian export volumes from 2003 onwards, compared to EU competitors such as Germany and France.

[10] Production prices on domestic markets have decreased more than those on foreign markets (5.4% versus −2.6%) and this anomaly can be explained by the recession's stronger impact in Europe than in other countries, so that the fall of domestic demand has induced European enterprises to suffer greater losses of profit than foreign ones.

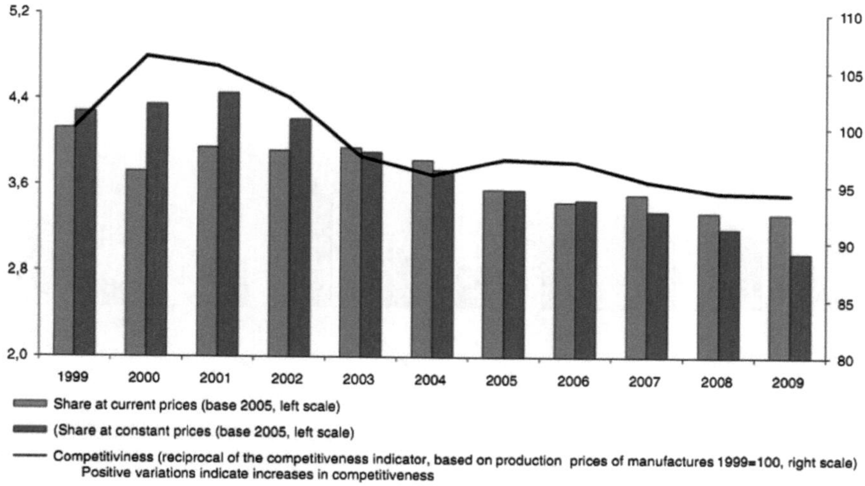

Fig. 2 Competitiveness and market shares of Italian exports (Source: ICE Report 2009–2010)

On the one hand, with the aim of escaping international competition, firms have improved the quality of their product, in terms of design, materials and components used. On the other hand, strong competition has triggered a kind of natural selection among firms, with the exit from the markets of companies focused on low quality goods, to the advantage of those of better quality. This has been verified by many traditional sectors of *made in Italy* (Bugamelli 2007), and many studies[11] confirm this behavioural strategy of Italian firms towards Chinese competition: comparing export shares of Italy and China in 2002 and 2005, for three different product types (footwear, clothing and furniture) in two European countries such as France and Germany, it emerges that Chinese goods have penalized our exports, but Italian firms have reacted by exporting goods with prices three times higher than the Chinese ones. Qualitative upgrading represents one of the possible strategies for responding to the challenge of the Chinese giant.

Exchange of Italy with its partners has been strongly affected by the fall of foreign trade in 2009. The upturn is taking place, but with different times and ways. Exchanges with emerging countries are rapidly growing (above all with China, India, other Asian countries, Mercosur and Turkey), but the recovery is slow with the United States, Japan and the European Union. The improvement of market shares compared to 2007 (in non-EU countries, Russia, North Africa, and in some Asiatic markets, China included) can be traced back to the exports of intermediate and investment goods, generated by the processes of internationalization of Italian firms. The geographical distribution of exports is still little oriented towards

[11] Cranec Report (2008).

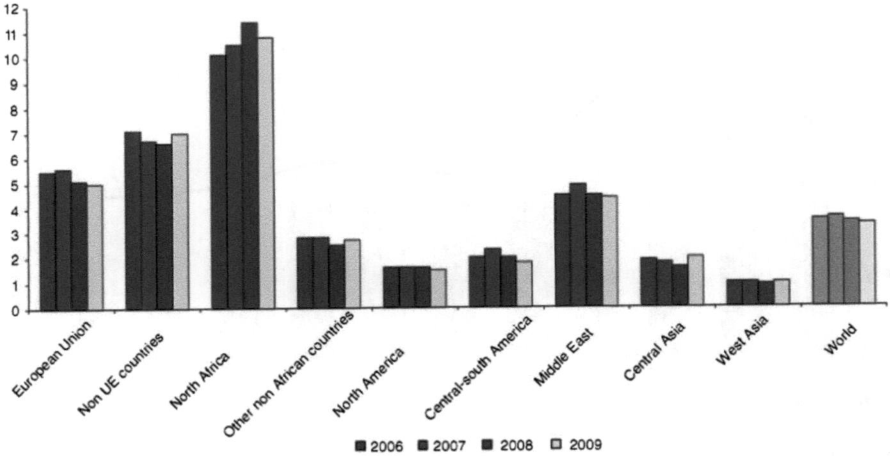

Fig. 3 Italy's market shares, per geographic area, at current prices (Source: ICE Report 2009–2010)

emerging Asian markets,[12] and this reduces the possible benefits related to recovery. The cause of this delay is the crowding out effect of Chinese exports, or rather the strong resemblance of consumer goods of traditional sectors with the local production of the same goods.

The crisis of 2008–2009 has had relevant effects on the sectoral distribution of exchange. The structural deficit of Italy in primary products has considerably decreased, thanks to the drop in price of raw materials, and this has offset the reduction of the manufacturing surplus. In 2010, the inversion of the cycle has again worsened the balance for primary products, while manufacturing surplus has kept falling because of the lack of necessary confidence that would allow the return to pre-crisis levels of exchange.

To reassert what has been said before about advantages and disadvantages, market share of exports has decreased in all sectors, but above all in footwear, textiles and furniture. The data (Fig. 3) confirm the evolution of the specialization model of the Italian economy, with the mechanical industry performing well, and the food sector, which has recovered having started from one of the lowest levels of the last 10 years.

Shifting from original production of final goods to the production of machinery for their production[13] translates into a rise of the productive chain, in favour of

[12] For instance, in East Asia the share of Italian export is 6% versus 7.9% of Germany and 7.4% of France.

[13] Such as machines for agriculture, machine tools, machines for packing, motors, etc. There are plenty of examples: in the footwear district of Vigevano they went from the production of shoes to making machines for the production of shoes.

intermediate and instrumental goods. This phenomenon, which is evident in machinery and food sectors, has concerned all the sectors of *made in Italy*, though in different ways (De Argangeli and Ferri 2005). It is interesting to note that inter-industry relationships that develop among machinery producers and final goods producers are full of innovation and technological improvements, along with the choice of the productive delocalization to exploit cost saving. According to some interpretations (Cranec Report 2008), such behavior represents an attempt by district firms to keep their competitiveness through processes of productive outsourcing that favour the export of machinery, necessary to the production of those goods.

3.2 Foreign Direct Investment

With direct foreign investment, enterprises acquire the control of productive activities abroad, and can do it either with the creation of new productive plants (greenfield), or through the acquisition of capital shareholding of the foreign company (brownfield).

For the period from 2001 to 2007, there was an increase in active internationalization, both in terms of participating foreign companies and sales. Italy is well positioned among traditional trading countries such as France and Germany, but also is strengthening its position among economies which are new EU members, such as Poland, the Czech Republic and Hungary. Other Eastern European countries, such as Romania and Bulgaria have become important targets for Italian firms, that have acquired capital shareholding in many productive units of small and medium size, specialized in production of *made in Italy*. Presence in the Middle East is oriented on the one hand towards sectors like machinery and means of transport, with a view to controlling sales markets, and on the other hand towards textile-clothing for which the low labor cost makes a difference.

However, the Isae Enquiry of 2010 suggests a clear about-turn of enterprises that have delocalized or are planning to do it in the near future. The crisis seems to have forced firms, especially the smallest ones, to fall back to their national territory. More optimistic evaluations come from the IMF,[14] according to which the recovery of FDI within 2011 will allow the 2008 level to be exceeded, thanks to factors such as the low labor cost in emerging countries and quality of work and infrastructure in industrialized countries.

If data on ownership of productive assets abroad exhibit a slow down in 2008, when compared to previous years, we have to point out that sales of foreign subsidiaries of Italian companies, still for the period 2002–2008 has increased more than export of

[14] It is the World Investment Prospects Survey 2009–2011 of FMI (2009).

goods and services. This seems to suggest that in the presence of a more mature strategy of internationalization, the choice has been that of replacing export with supply from the foreign affiliated companies (Ice Report 2010).

FDI show positive effects not only on the growth of the country of origin but also on the destination one. Incoming FDI can be assumed as the attractiveness of a territory that can generate wealth, and allow the introduction of technologies and knowledge, whereas outgoing FDI measures the capacity of a country to expand its presence abroad. Istat data[15] show that Italy is far behind other countries, such as France and Germany, as far as incoming FDI are concerned. This has to do with the sectorial composition of investments that prefer productions with strong economies of scale and a high intensity of research, which are not so common in Italy.

In general, the majority of Italian companies that are investing are part of a larger size class, and this is confirmed by data of 2008: three quarters of employees and 87% of Italian subsidiaries refer to mother companies with more than 250 employees. This empirical evidence outlines the relationship between size, complexity of organization and investment capacity abroad, but it is important to underline that the most dynamic growth of productive investments leaving Italy has to be ascribed to companies of medium size, that is between 50 and 249 employees.

This positive performance deals with the characteristics of the Italian productive model: the lack of big multinational companies, but the presence of a productive tissue made of a multitude of small and medium enterprises has led the medium firms in particular to play a dominant role in the process of internationalization, at relatively high costs that, for small firms, represent a disincentive to going abroad.[16]

Geographical distribution of capital shareholding by employee class (Table 1), suggests once again the relationship between firm size and distance from the markets, which we also find to be the case for exports.

The role of micro-enterprises is marginal and is limited to a presence in the markets closest to home. Subsidiaries of small investors have produced in 2008 3.7% of turnover, even being 6.7% of employees, with a slight decrease compared with 2003. However, it is appropriate to underline that these data represent an underestimate of internationalization activities of small and medium enterprises that, as said before, make use of lighter modes (commercial or technical-productive agreements) to have more flexibility and to bear lower fixed costs. This aspect has to be taken into consideration while evaluating their role in the internationalization process.

[15] It is Istat enquiry on structure and activities of Italian subsidiaries, resident in Italy (Inward FATS), published in 2010, for the year 2008.

[16] These enterprises, the medium size, have been defined in literature the fourth capitalism (Mariotti and Mulinelli 2008).

Table 1 Distribution (in %) of employees of foreign subsidiaries by geographic area of localization and size class (employees) of the investing firm

Classes of employees	From 1 to 49	From 50 to 249	250 and over	Total	Weight of the area (in %)
	2003				
European Union	7.7	18.7	73.6	100	52.4
Non EU countries	8.3	15.8	75.9	100	10.9
Middle East and North Africa	20.4	25.3	54.3	100	3.3
Other African countries	7.7	11.4	90.9	100	2.5
North America	4	10.3	85.7	100	7.6
Central-South America	2.6	10	87.4	100	13.2
Central Asia	7.4	20.8	71.8	100	2.2
East Asia	7.2	18.7	74.1	100	7.2
Oceania	3.3	9.9	86.8	100	0.8
Total	7.1	16.6	76.3	100	100
	2008				
European Union	7.6	19.5	72.9	100	51.1
Non EU countries	6.1	14	79.8	100	13
Middle East and Northern Africa	17.4	19.4	63.3	100	3.5
Other African countries	3.5	14.3	82.1	100	2.1
North America	3.5	17.5	78.9	100	7.2
Central-South America	2.7	11.7	85.5	100	11.8
Central Asia	7.7	23.9	68.5	100	2.2
East Asia	6.6	20.9	72.5	100	8.6
Oceania	4.9	16	79.1	100	0.5
Total	6.7	17.8	75.5	100	100

Source: ICE Report (2009–2010)

3.3 *Delocalization of Manufacturing Sector*

The main source on delocalization aspects of the manufacturing sector comes from ISAE inquiry, whose results are given in ICE Report of 2009–2010.[17] The term delocalization refers to the phenomenon offshore in-house sourcing, that is the shifting of a part of productive activity abroad, in a company of the same group. Still in literature we do not have a unanimous definition of off-shoring and productive internationalization (Crinò 2009).

Data show that from 2005, the phenomenon has been relevant for a minor part of the sector, i.e. 4%, halting in 2010. The Italian enterprises that later and more timidly have accessed the international context, have difficulty in adopting such a strategy as a sustainable path for dimensional growth, and this can be read also in

[17] The sample is of about 3,800 enterprises and covers 4% of the relevant universe, represented by Italian firms with at least 10 employees, as is reported by ASIA archive of Istat. The enterprises are asked whether they have decentralized a part of their production or whether they intend to do so within the next 12 months.

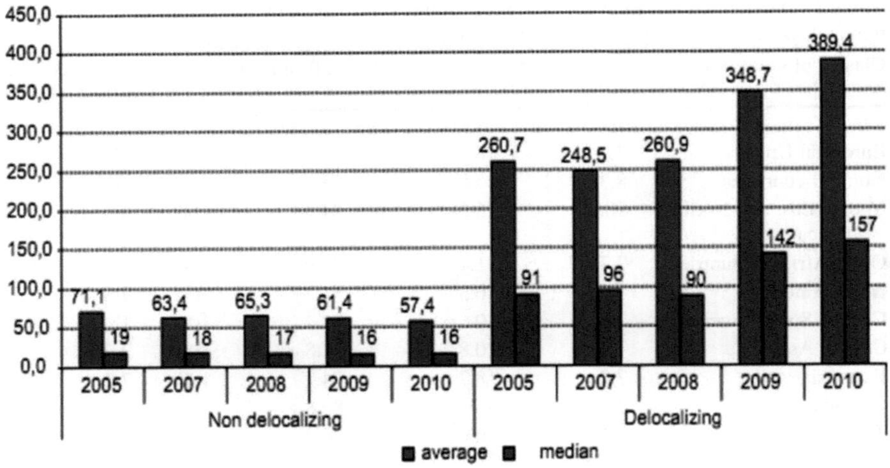

Fig. 4 Average size of delocalizing and non delocalizing enterprises (employees) (Source: ICE Report 2009–2010)

data that reveal the intention to delocalize in the next 12 months, with an almost insignificant result in 2010. However, besides the unusual economic situation it has to be highlighted that delocalization will interest the manufacturing section in a limited way.

The interested enterprises are the big ones, with a clear increase of the average size (Fig. 4), and a loss of ground by the small ones. The crisis has somehow accentuated a contraction of the phenomenon from 2008 until now in traditional sectors of *made in Italy* (textiles, clothing and leather) and in lighter way in high technology sectors (communications, medical and precision equipment).[18,19]

When dealing with destination areas, the relevance of the size enterprise reemerges, so that the bigger firms delocalize in markets further afield. China today represents a well-established destination for big, small and medium enterprises, whose presence has strengthened over time, from 17% and 25% in 2007 to 30% and 35% in 2010. Among destination areas, in first place we have Romania and east European countries, thus confirming a model of delocalization that gives priority to the curbing of production costs, rather than the defense of markets.

More suggestions come from data relative to the turnover of Italian enterprises that delocalize.[20] These values are lower for large companies, whereas more than

---

[18] Particularly in these sectors the few positive balances are recorded, between the share of enterprises that delocalize for the first time and the percentage of those with no productive capacity abroad: 5.71 and 1.79 respectively.

[19] This confirms what has been highlighted by Micelli et al. (2003): at the beginning low value added and labour intensive activities were transferred, but over the years small and medium enterprises have delocalized activities with a higher technology content.

[20] The data refer to the percentage of firms that delocalized for the first time, and those that did so previously, in 2008, 2009, 2010.

half of small ones in 2009–2010 no longer have productive capacity abroad. Therefore coherently with what has been said before, it is the small sized enterprises that are backing down in the activity of internationalization. As far as destination areas are concerned, China is the country showing more stabilization; for the other areas, the farther away they are, the more revealing is the defection: within this picture we notice the preferred destinations of Italian delocalization for Romania and Eastern Europe.

The analysis of the phenomenon for large district companies shows a similar result. Enterprises belonging to a district are usually bigger than those that do not, but this does not make a difference to the balance between newly delocalizing firms and those who did it previously, that for 2010 is as negative as for the complete aggregate. Belonging to the district however gives more stability to the delocalization activity and a better performance in general, so that fewer firms defect from productive internationalization.

Reasons behind delocalization trends of north east district SMEs are pinpointed in a work by Turato (2002) that, in order of importance, contains the availability of labor factor, the possibility of exploiting the final markets in the foreign country along with accessibility, due to geographic proximity and flexibility, in the sense of less restrictive legislation and custom controls.

Other studies assert the importance of delocalization in enterprises projects, according to which more than 30% of SMEs that delocalize, achieve abroad a share of total turnover higher than 50% (Capitalia 2005). The quantitative evolution goes along with a qualitative change, so that in the early stages firms transfer simpler productive steps, and in due course more complex ones, then eventually the entire productive cycle (with the exception of the stages up and down the manufacture). This is taking place together with an improvement of the quality of foreign subcontractors, through learning processes similar to those experienced by many Italian subcontractors. Therefore if originally low quality production is delocalized, then goods of medium and high quality are transferred. Delocalization becomes part of a more general strategy of reorganization on a global scale, and in some cases goes along with the competitive positioning towards high quality products, typical of many district firms specialized in *made in Italy* (Prota and Viesti 2007).

3.4 The Other Modes of Internationalization

Manufacturing enterprises often resort to modes of commercial and productive internationalization which are non-equity in nature.

As far as commerce goes, some actions have a determinant role above all for some sections of *made in Italy* (such as the setting up of show-rooms, the strengthening relationships with local distributors and importers, franchising agreements) in order to control markets better.

As far as production is concerned, it takes the forms of contracting and subcontracting agreements, joint ventures and usually relationships of non-equity cooperation, with enterprises operating in foreign countries.[21]

These modes of weak internationalization are anyway important in the growth strategies of firms, although available information to this regard is lacking.[22]

Even with modest organizing and operational structures, a part of SME has managed to be present in world markets, learning from the stock of specialist knowledge, accumulated over time. The small Italian entrepreneurship which has become international, often developed from "historical accidents", in Krugman words, very heterogeneous, and meanwhile has been able to enter the global value chain, thanks to the export of a "process philosophy", geared towards quality and the use of firms' networks. This *made in Italy*, which is not so well-known, owes its success to a productive and cultural heritage which is not replicable, and experiences a "gentle" globalization, which is "weak" at the same time.[23] Gentle for its capacity to raise, mutually, the value of competitive assets of commercial and productive partners, weak for the (structural) incapacity to fully control enterprise networks, that allow the firm to internationalize. And even with basic differences in the typical modes of large multinational companies, this modus operandi of small and medium enterprises allows operational synergies to be released and cuts transaction costs due to entry into new markets, especially when there is the need for forging new commercial and productive relationships.

4 International Integration of Productive Chains

To adequately measure the level of internationalization of a productive system like the Italian one, it is necessary to evaluate market and cooperation relationships that establish an economic link, not based on ownership. In recent years, for many manufacturing firms it has been profitable to adopt policies of trade intermediation, that expect to acquire at low prices from producers whose currency has depreciated, with the subsequent sale of final products. Import and export data can give interesting suggestions on the direction and intensity of recourse to foreign subcontractors, as from the 1990s, a remarkable increase in imports has taken place, from emerging

[21] A study by Centro Studi Unioncamere-Assocamere Estero (2002) highlights that subcontracting is one of the major strategies of district firms, even preferred to FDI and *joint-ventures*, due to the high fixed costs associated with these last categories.

[22] In research by Formez in 2004, out of 764 small and medium sized enterprises, only 12% used forms of equity internationalization, while more than 40% had achieved stable contractual relationships with foreign enterprises outside the EU, or had used subcontracting with operators usually located in Eastern European countries.

[23] This work by Bonomi, in Unicredit Report 2010 on small enterprises is based on declarations collected in 2009/2010 from a group of excellent enterprises located on different productive platforms of the country.

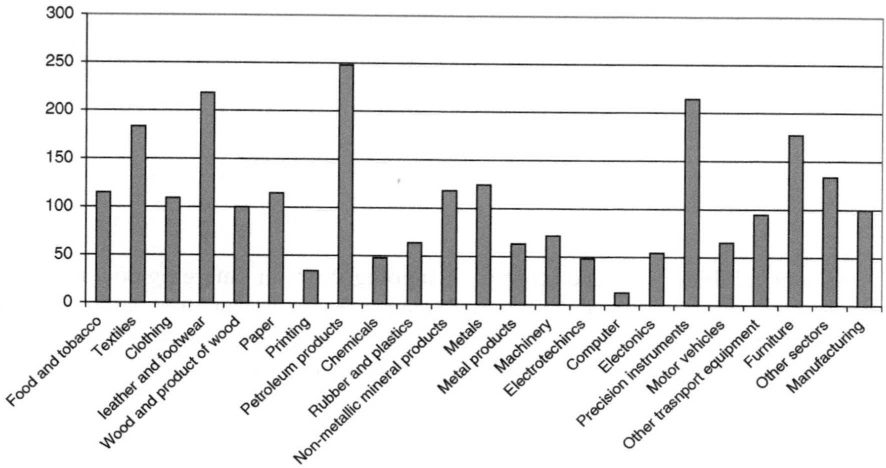

Fig. 5 Capacity of control of Italian manufacturing industry (average = 100) (Source: Own elaborations on data from Tattara et al. (2006))

countries, particularly in areas with a strong productive specialization.[24] The indicator proposed by Foresti and Trenti captures the capacity of Italian firms to control import flows from emerging countries and is directly influenced, although it is not a direct measure, by phenomena of productive relocation that expect the partial transfer of the productive cycle and the necessity of effecting one of the final phases of transformation in the Italian territory. The registered values are high when imports coming from low labour cost areas are directed towards areas that are specialized and strong importers of the same goods which are exported, whereas they are negatively influenced by the amount of exports from emerging countries that do not pass through those areas specialized in imported goods. Whilst not a real measure of internationalization (not all situations are recorded, as when goods produced abroad by an Italian firm do not pass through the province where the Italian firm is located, but are directly exported from another country), gives interesting indications as to the level of control of different manufacturing sectors.

Moderate control (Fig. 5), below average values of the manufacturing industry can be seen for the chemicals sector, rubber and plastics, instrumental machinery, computer, electronics, electrotechnics and motor vehicle sectors. The latter experience other modes of internationalization (also FDI). Outsourcing and productive delocalization in metals and machinery sectors becomes less relevant as we go down the chain, because the intensity of the use of labour decreases and costs are growing with fragmentation of production in complex systems like machinery.

[24] A significant example is footwear, where imports from emerging countries (China and Romania above all) covered 77% of total import in 2004, for the Italian provinces where the main footwear districts are located (Foresti and Trenti in Tattarà, Corò, Volpe 2006).

Table 2 Innovative activities of manufacturing firms, according to size (values in %)

Firm size	Product innovation	Patents	Total firms
Micro firms (0–9)	21.9	5.2	83
Small firms (10–49)	5.4	2.2	15.1
Medium firms (50–249)	0.7	0.5	1.9
Total	28	7.9	100

Source: Pittiglio et al. (2009)

Sectors exhibiting a higher control of ingoing flows in emerging countries are those of fashion (textile and clothing, leather, footwear and optical products) petroleum and food products. The remarkable values of the fashion system highlight the intensity of the trade relationship between Italian firms and those located in emerging countries through outsourcing of labour intensive steps of production and purchase of final goods in order to enlarge the range of supply. Also the furniture sector demonstrates higher than average control that however has reduced over time. The evolution of the index indicates that an increasing share of furniture imports from emerging countries is not handled by the Italian firms producing furniture, even though an intensification of relocation activity has taken place.

5 Some Empirical Evidence on the Relationship Between Innovation and Internationalization

Significant research has been conducted by Pittiglio, Sica e Villa (2009) on a *dataset* on small and medium enterprises of manufacturing sectors, for the period 2001–2003.[25] Data (Table 2) show that 28% of the sample group achieved product innovation, and 8% registered at least a patent in order to protect their innovation.

These low percentages of innovative activities reflect some strong peculiarities of our productive system, characterized by a high presence of micro firms (83% of firms in the sample have less than nine employees) and a marked specialization in traditional sectors. Furthermore, if we consider geographical distribution, a major part of innovative firms are concentrated in the north of the country, while only the 21% is located in the south, and 70% of these are micro firms. The situation is not different when we consider patents as proxy of innovation (southern firms are only 16%, vs 55% of northern firms), and this touches on an age-old question of the big gap between north and south, on economic grounds, among others.

A different mode of internationalization (exports a/o agreements) produces a different result: among exporting firms 44% has achieved product innovation, and 23% has registered a patent; those with a cooperation agreement with foreign firms perform better, both for product innovation (50%) and for registration of a patent (31%).

[25] 2,603 enterprises were considered. The level of response (owners and managers were interviewed) was 76%: quite high if compared to similar studies in England, with lower rates of 42%.

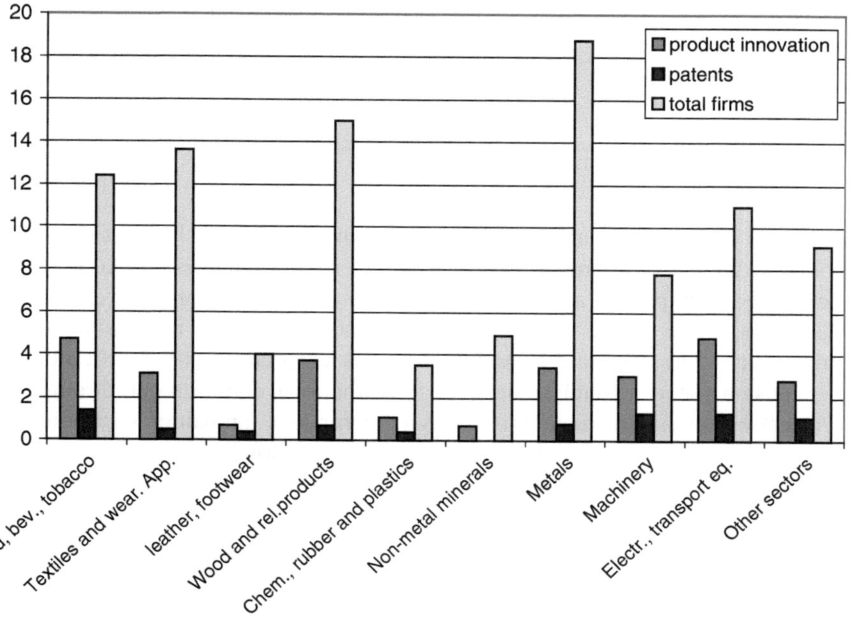

Fig. 6 Sectorial distribution of innovating activities for Italian manufacturing firms (Source: Own elaborations on data by Pittiglio et al. (2009))

The relevant markets are those of EU-15, United States and Japan. The preliminary results confirm that firms globally engaged through cooperation agreements are able to generate more innovation compared to their colleagues, located in the domestic markets, whose percentages are definitely low (23.6% e 6.7% for product and patent innovation).[26]

The next step is to consider the sectorial distribution of innovation. Figure 6 demonstrates that most innovative firms are concentrated in sectors of electronics and transport equipment (4.8%), followed by food, beverages and tobacco (3.7%). When patents are considered, sectors do not change except for rank, and with lower values. Finally Table 3 gives information on the three knowledge inputs: innovation expenditure, internal and external transfer of knowledge. Internationalized firms make greater use of inputs, in order to produce new ideas, and this is verified for all inputs concerned. In particular, these firms use external transfers of knowledge to a greater extent, if compared with domestic firms (29.9% versus 19.2%), and this is proving to be a favoured channel for diffusion of technological knowledge.

---

[26] A similar result was reported by the Unicredit Report on small firms (2010), according to which internationalized firms are usually more innovative (51.1% of those selling abroad, versus 33.6% of those operating on the domestic market). Data refer to firms that have undertaken innovation or significant changes in their activities within the last 2 years.

Table 3 Use of knowledge inputs, for international and domestic firms (values in %)

Knowledge inputs	International firms	Domestic firms	Total
Innovation expenditure	69.0	44.3	49.7
Internal flows	40.3	23.5	27.2
External flows	29.9	19.2	21.6

Source: Pittiglio et al. (2009)

In the choice of variables for the econometric estimation Pittiglio et al. (2009) follow a subjective approach, or rather, the observation unit is the product/process developed or improved at firm level: firms are classified as innovative if they have been innovative in the last 2 years, not-innovative if no innovation has been undertaken. The innovative output, in terms of product and patent innovation depends on a set of variables: the level of international commitment (through export a/o collaboration agreement with foreign enterprises), innovation expenditure, and the amount of internal and external transfer of knowledge. The results of the estimation confirm what has been outlined above in literature reviews and are in line with evidence of other recent studies for UK and Germany. Firms globally engaged generate more knowledge than those operating solely in the domestic market, and this is true for both the measures that define the innovative output. This can be explained by a higher expenditure in innovation, but it is the access to larger flows of ideas from external channels that makes the difference. In other words, firms have a better opportunity to acquire knowledge through suggestions and proposals from people involved in the productive process, existing links between firms and its competitors or external organizations (such as universities, research centres, scientific and technological parks).

Hence, to generate innovation, a conspicuous investment in innovation has to be supported by international commitment and by the appropriate use of external sources of knowledge and information.

6 Innovation and Productivity

Some recent empirical studies have pointed out that firms with better productivity tend to self-select in terms of international presence (Tybout 2003), and this has boosted a whole series of models that expressly try to formalize firm heterogeneity.[27] Despite scant empirical evidence on export, there are many contributions documenting that international firms tend to have a definitely better performance compared to those not engaged in foreign investment.[28]

[27] See the contribution of Bernard et al. 2003.

[28] The relevant contributions for country are: Barba Navaretti and Castellani (2004) for Italy, Criscuolo and Martin (2003) for UK, Pfarffermayer and Bellak (2002) for Austria.

However, reasons behind firm heterogeneity have not been investigated. As Zanfei and Castellani (2007) underlines, the levels of productivity are casually drawn from a probability distribution, and firm behaviour varies with a given level of fixed costs and negotiation, for operating abroad. This kind of hypothesis ignores factors at the base of heterogeneity, and does not explain why firms differ one from the other.

A recent strand of literature has tried to open the black box. Castellani e Zanfei (2007)[29] highlight that Italian firms that are seriously committed on the international front, are characterized by higher productivity premiums and by better innovative performances. R&D activities are significantly responsible for firm heterogeneity. More specifically, firms with a minor presence on foreign markets (those basically controlling sales and not production) exhibit a level of productivity that lies between that of firms that manufacture abroad and firms that export, without demonstrating a higher level of innovation than the latter. The indications in terms of policy favour the stimulation of firms with a presence abroad, because technical growth and research activities carried on *intra-muros* are enhanced.

Using a large sample of Italian manufacturing firms for the period 1998–2003 Castellani and Giovannetti (2010) make the point that heterogeneity in total factor productivity is related to a better capacity of innovation and use of knowledge workers (devoted to R&D, managers and clerks), Also international firms exhibit superior organizational and managerial practices. Lopez (2009) with plant-level data for Chile finds that firms increase their productivity before they begin to export, so confirming the idea that to enter foreign markets a sort of self-selection is active.

The work of Antonietti and Cainelli (2010) goes in the same direction: they endogenize firm heterogeneity and make it dependent on innovation that in turn relies on the decision to invest and on the amount of investment devoted to innovative activities. Descriptive statistics based on the considered database[30] confirm that exporting firms are larger, more productive and make more profits compared to the non exporters. The same can be said for the innovative ones, that is, firms that have achieved at least one product a/o process innovation (Table 4). Estimates produce interesting results concerning the relationship between productivity and exports, that seem to influence each other: the most productive firms have a better performance for export, and those with better results on international markets are also the most productive. The impact of productivity on export is

[29] The *dataset* used refer to the Second Community Innovation Survey (CIS-2) and to European Linkages and Ownership Structure (ELIOS); 778 firms were considered and the period during which an innovation was undertaken goes from 1994 to 1996.

[30] It is the database built on the IX Enquiry into manufacturing firms, conducted by Unicredit-Capitalia (ex Mediocredito Centrale). Interviews were conducted in 2004 to a representative sample of firms active between 2001 and 2003. All the firms with more than 500 employees are included in the dataset, the sample with between 11 and 500 employees has been stratified by localization and industry.

Table 4 Statistics on innovation, productivity and export

Average 2001–2003	Exporters	Non exporters
Sales per employee	255.65	244.75
Value added per employee	54.22	50.9
Size (n. of employee)	135	76
Average 2001–2003	Innovating	Non innovating
Sales per employee	242.53	265.99
Value added per employee	54.26	51.07
Size (no. of employee)	125	103
Export (per employee 2003)	100.79	97.41

Source: Antonietti e Cainelli (2010)

significant, and this strengthens the idea that the two hypotheses of self-selection of firms (i.e. the ability of the firm determines both its presence on the international market and its innovative performance) and of *learning by exporting* (i.e. export activity leads the firm to improve its products or to introduce new ones), can coexist and that one does not exclude the other.

7 Concluding Remarks

The Italian economy has been caught by the crisis in a condition of structural fragility attributable to fundamental problems that undermine the recovery of productivity, such as the degree of openness and competitiveness of markets, fragmentation of the productive fabric, the small size of firms, inadequacy of public policies for material and immaterial infrastructures of development. If before the crisis of 2008/2009 exports represented a large proportion of aggregate demand, today we expect from it a fundamental role, and just firms of the so called fourth capitalism are those to have increased their contribution to exports, over the previous 10 years. However the competitive context has now changed and the protagonist countries try to cover a key position in global chains.

The opening and the fragmentation of chains and the repositioning of comparative advantages represent complex phenomena that interact with each other and have effects on evolution of trade flows which are often not clear. Where FDIs are concerned, our country remains behind our principal European partners, but a leading role is covered by Italian firms in the process of redefining international divisions of labor between old industrialized and emerging countries. Even with different evolutions and performances, it is worth looking at some behaviour patterns.

First, it seems to be emerging that Italy goes from being a country specialized in finished goods to a country that is positioning itself up the productive chain, with the production of intermediate goods that are used by other countries, while keeping a competitive advantage in the production of investment goods.

The so called traditional sectors are those that have been hit hardest by the mentioned phenomena (mainly the fashion sector, but also the home sector (furniture, household appliances, tiles)). The decline of comparative advantages in this context means a relevant loss of trade share abroad and a significant increase of foreign export penetration. Through new management methods and organization of the productive chain, Italian firms have structured chains of international production on the basis of product characteristics and relationships, making direct use of more hierarchical forms such as FDIs, whenever this was required by the complexity of the product and high transport costs.

The emerging picture sheds both light and shade, but is essentially encouraging. If many micro firms have been blown away by the crisis (above all when decentralization has been achieved as an economic strategy, or even as a diffusion of economic activity outside the large enterprise, without sizeable growth), for many others the process of qualitative upgrading of products, as well as the strategic one, is proceeding. The latter is based on a marketing mix approach with high flexibility and speed, to better gather opportunities from global markets (Rapporto Unicredit 2010).

From 2001 to 2007, active participation has increased both in terms of subsidiaries and turnover. However 2010 data show a reversal of this trend, with a clear about-turn of firms that have delocalized or foresee doing so in the near future. The crisis has led enterprises, mainly those smaller ones, to turn back towards the domestic territory. The stronger growth of outgoing productive investments can be traced back to medium sized enterprises, and this confirms the close relationship between size and complex organization of activities. Medium enterprises play a leading role in the internationalization process, bearing costs that for small firms represent an insuperable impediment to entering foreign markets.

Manufacturing enterprises often resort to modes of commercial and productive internationalization which are non-equity in nature, weak but relevant modes in the growth strategy of firms. Such globalization is defined as soft, because who takes part in the process is able to enhance his own competitive assets, but weak at the same time because there is no full control of networks that allow the firm to internationalize.

One of the big issues of our industrial system is the modest propensity to innovate in small and medium enterprises. Statistics relative to the number of patents, researchers and expenditure share on R&D, position Italy in the lowest place in European rankings. Empirical evidence for Italy suggests low innovation as far as SMEs are concerned, attributable to peculiarities in our productive system, like the presence of numerous micro-firms and a specialization in traditional sectors. The mode of internationalization makes a difference: enterprises with collaboration agreements are able to generate more innovation than their colleagues localized in national markets. Sectorial distribution of innovation reveals that the majority of innovative firms concentrates on electronics and transport equipment, followed by food products, beverages and tobacco.

Those active at international level make greater use of input such as innovation expenditure, internal and external transfer of knowledge and are able to generate new knowledge. This can be explained by higher expenditure on innovation, but access to information flows through external channels is essential, and therefore is the best opportunity to acquire knowledge through suggestions from operators involved in the productive process (customers, suppliers), existing links between the firm and its competitors or external organizations (such as universities, research centres, scientific and technological parks).

However, if innovation represents a competitive factor that affects technology, then new products and processes, design, marketing, small and medium firms in districts can also achieve this activity. Statistics often do not gather the extent of innovation that not only comprises the intensity of R&D and number of patents, but also includes the sending of experienced technicians for after-sale assistance: in this way firms achieve "black" innovation, and goods and services produced become more complex because more technology is involved. Hence the problem of innovation in Italy is that there are no big groups investing considerable amounts in research and patenting, or rather, large companies able to take on formal research versus informal research of small and medium firms.

The size constraint typical of our productive system represents a structural problem that influences the delay in labour productivity and investment growth. In order to export, it is necessary to be larger and more productive, with a major use of capital and skilled labour. There is a minimum threshold below which it is very hard to bear the costs and innovations necessary to compete on global markets. The small dimension implies at the same time a polarization onto a limited number of markets, with strategy focused on one single area, usually within old Europe. The firm profile can be different according the duration of activity abroad: those recently internationalized are more oriented towards a single market, those with more international experience accomplish geographic diversification, and this confirms that growth on the path to internationalization is also possible for small firms.

To pursue dimensional growth in existing firms, at all costs, with the idea that automatically this strategy will bring new innovating abilities and therefore a better perspective of internationalization, is misleading. The evolutionary trends of enterprises before the crisis, in terms of qualitative improvement, should be boosted in order that they completely take into account inputs coming from markets, in terms of technological and organizational change. Growth in size at this point will be a natural consequence, and could trigger the virtuous circle of productivity growth and the process of internationalization, two factors that according to literature strongly influence each other.

Difficulties due to their small size can be overcome through firm networks, fundamental for acquiring that critical mass necessary to sustain over time a competitive position on global markets. All the same, the level of collaboration is still very low, often overwhelmed by local relations and by problems of confidence. The form of historical aggregation, districts, today is going through a phase of profound change, and has been flanked by that of global chains, through which small enterprises can address global markets, managing to contain to the utmost the high fixed costs generally associated. The ability to exploit the benefits that arise from network participation depends crucially upon a system of rules, incentives and infrastructures that can stimulate production and promote long-term growth.

Finally, the major inefficiencies of the country have to be considered as constraints to firms in international markets, as well as the possible role of the public system in supporting firms along this route. The most serious and age-old problem is of course that of public debt, a heavy burden that absorbs valuable resources which otherwise could be assigned to infrastructural investment. To which is added, on the one hand the question of the energy deficit, which is a huge drain on our trade balance, given the strong dependency of the country on petrol and gas, and on the other hand the problem of the marked territorial divide between north and south which is worsening rather than improving. As a matter of fact, Mezzogiorno contributes insignificantly to export growth (Rapporto Cranec 2008).

In view of the above, it is vital for firms to have a system of public support that, through a global strategy of economic policy, might focus on innovation and internationalization. This should lead to a series of regulations, financial interventions and measures (in this regard the role of the banking system is crucial) to foster the working out and consolidation of firms' international positions, and help them to overcome the high initial costs of accessing foreign markets.

The international openness of a country can be enhanced through public action, at different institutional levels, starting from trade policies at community level, to those for competition, which are vital for market dynamism, to those for international cooperation. The presence of internationalized and well organized firms has the prerogative of strong public investment in the formation of human capital and research, as well as in the enhancement and upgrading of specific territorial assets. Only then the Italian model of specialization can renew itself and push the country forward towards economic growth that will be able to combine economic progress with a civil and institutional one.

References

Antonietti R, Cainelli G (2010) Innovation, productivity and export evidence from Italy. Openloc working paper no 17

Centro Studi Unioncamere – Assocamere Estero (2002) Internazionalizzazione, imprese e territorio. II° Rapporto, Roma

Barba Navaretti G, Castellani D (2004) Does investing abroad affect performance at home? Comparing Italian Multinational and National Enterprises, CEPR Discussion Papers

Bernard A, Eaton J, Bradford JJ, Kortum S (2003) Plants and productivity in international trade. Am Econ Rev 93(4):1268–1290

Bugamelli M (2001) Il modello di Specializzazione Internazionale dell'Area dell'Euro e dei Principali Paesi Europei. Omogeneità e Convergenza. Banca d'Italia, Roma

Bugamelli M (2007) Prezzi alle Esportazioni, Qualità dei Prodotti e Caratteristiche d'impresa: un'Analisi su un Campione di Imprese Italiane. Temi di Discussione, No. 634, Banca d'Italia

Capitalia, (2005) Indagine sulle imprese italiane. Rapporto sul sistema produttivo e sulla politica industriale. In Osservatorio sulle piccole e medie imprese, ottobre, Roma

Castellani D, Giovannetti G (2010) Productivity and the international firm: dissecting heterogeneity. J Econ Policy Reform 13:25–42

Castellani D, Zanfei A (2007) Internationalization, innovation and productivity: how do firms differ in Italy? World Econ 30(1):156–176

Centro Studi Unioncamere - Assocamere Estero (2002), Internazionalizzazione, imprese e territorio, II° Rapporto, Roma

Cranec (2008) L'Internazionalizzazione del Sistema Industriale Italiano, una Sfida Vincente delle PMI e dei Distretti Italiani, Rapporto 2008

Crestanello P, Dalla Libera PE (2003) La Delocalizzazione Produttiva all'Estero nell'Industria della Moda: il Caso di Vicenza. Econ Soc Regionale 21(2):5–46

Crinò R (2009) Offshoring, multinationals and labour market: a review of the empirical literature. J Econ Surv 23(2):197–249

Criscuolo C., Martin R. (2003) Multinationals and U. S. Productivity Leadership: Evidence from Great Britain 107:134–149.

De Argangeli G, Ferri G (2005) La Specializzazione dei Distretti: dai Beni Finali ai Macchinari del Made in Italy. In: Signorini LF, Omiccioli N (eds) Economie locali e competizione globale. il Mulino, Bologna

De Nardis S (1997) Fenomeni di Persistenza e Cambiamento nelle Specializzazioni dei Paesi Industriali. Riv Polit Econ 87(1):89–105

FMI (2009) World investment prospects survey 2009–2011

Foresti G, Trenti S (2006) Apertura delle filiere produttive: la nuova collocazione dell'industria italiana nello scenario internazionale. In: Tattara G, Corò G, Volpe M (eds) Andarsene per continuare a crescere. La delocalizzazione internazionale come strategia competitiva. Carocci, Roma

Formez (2004), La governance dell'internazionalizzazione produttiva. L'Osservatorio. A cura di A Rossi, A Cenderello, Quaderno n. 28, Roma

Gereffi G (1994) The organization of buyer-driven global commodity chain: how US retailers shape overseas production networks. In: Gereffi G, Korzeniewicz M (eds) Commodity chain and global capitalism. Praeger, Westport, pp 95–122

ICE (2010) Rapporto 2009–2010

Jones R, Kierzkowski H (2000) A framework for fragmentation. In: Arndt S, Kierzkowski H (eds) Fragmentation and international trade. Oxford University Press

Lopez RA (2009) Do firms increase productivity in order to become exporters. Oxford Bull Econ Stat 71(5):621–642

Mariotti M, Mulinelli S (2008), Italia Multinazionale 2006. Le partecipazioni italiane all'estero e estere in Italia. Rubettino Editore.

Micelli S, Chiarvesio M, Di Maria E, (2003) Processi di internazionalizzazione e strategie delle imprese distrettuali tra delocalizzazione e innovazione. Paper presentato alla Conferenza ICE: Internazionalizzazione e frammentazione della produzione dei distretti industriali, Roma, 20–21 Marzo

Onodera O (2008) Trade and innovation project a synthesis paper. OCSE trade policy working paper no 72

Pfaffermayer M, Bellak C (2002) Why foreign-owned are different: a conceptual framework and empirical evidence for Austria. In: Juncnickel R (ed) Foreign-owned firms: are they different? Palgrave-Macmillan, Houndmills

Pittiglio R, Sica E, Villa S (2009) Innovation and internationalization: the case of Italy. J Technol Transfer Springer 34(6):588–602

Prota F, Viesti G (2007) La Delocalizzazione Internazionale del Made in Italy. L'Industria, luglio-settembre

Tattara G, Corò G, Volpe M (2006) Andarsene per continuare a crescere. La delocalizzazione internazionale come strategia competitiva. Carocci, Milano

Turato F (2002) Il Ruolo delle PMI del Nord Est nei Balcani. Collana Ricerche, no. 8, Fondazione Nord Est, Venezia

Tybout JR (2003) Plant and firm-level evidence on 'new' trade theories. In: Harrigan J, Kwan Choi E (eds) Handbook of international trade. Blackwell, Basil

Unicredit (2010) Rapporto Unicredit sulle piccole imprese, VII Edizione, 2010–2011

Venables A (1999) Fragmentation and multinational production. Eur Econ Rev 43:935–945

The Great Recession of 2008–2009, Conventional and Non-conventional U.S. Federal Government Responses and Their Impact on U.S. Small Businesses

Giuseppe Gramigna

1 Introduction

This paper provides a general background on the events that lead to the financial crisis and Great Recession of 2008–2009 in the United States. A strong argument can be made that the origins of the crisis can be traced to a fundamental imbalance between house prices and personal income. House prices outpaced personal income, as a result mortgage payments took an even bigger portion of disposable income, and the financial sector became evermore concentrated on residential investment. Around 2007, however, U.S. financial markets began to question the valuations of financial products based on residential mortgages. This resulted in a pronounced drop of inter-bank confidence, leading to a dramatic rise in the relative cost of capital for financial institutions, and a sharp reduction in lending. As a result, the economy experienced a pronounced recession, hitting small businesses particularly hard.

The U.S. federal government responded with enormous conventional and non-conventional measures to assist the financial markets and stimulate the economy. In addition, the federal government undertook numerous actions to address the particularly sharp drop in the small business sector by providing additional incentives to (1) stimulate the small business lending supply, and (2) fortify small businesses' balance sheets in order to increase small business loan demand. The U.S. Small Business Administration (SBA) played a critical role in increasing access to capital for small businesses.

The financial and economic crisis had a pronounced impact on the SBA's Capital Access Programs, with the volume of its two largest loan guarantee programs experiencing an approximate 60% decline. The SBA addressed these challenges by providing additional incentives to financial institutions and by assisting in the unfreezing of the secondary market for SBA loans.

G. Gramigna (✉)
U.S. Small Business Administration, 409 third Street S.W, Washington, DC 20416, USA
e-mail: giuseppe.gramigna@sba.gov

G. Calcagnini and I. Favaretto (eds.), *Small Businesses in the Aftermath of the Crisis*, Contributions to Economics, DOI 10.1007/978-3-7908-2852-8_8,

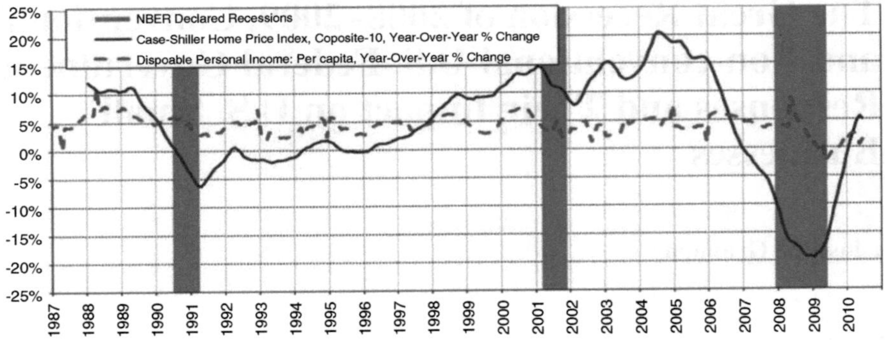

Fig. 1 House prices and personal disposable income (Source: Standard & Poor's (2010), S&P Case-Shiller Home Price Index, U.S. Department of Commerce, Bureau of Economic Analysis (2010), NIPA, and SBA calculations

2 The Origins of the Crisis

A cursory look at the data for *house prices* and *per capita personal disposable income*, shown in Fig. 1, reveals that during the 1998–2007 period, house prices outpaced increases in per capita personal disposable income by more than 1% point. More importantly, during the 2003–2006 period, house prices increased 13%, while per capital personal disposable income grew 4.5%.[1]

As a result, housing took a bigger portion of the consumer's income. For example, in 1996 the median house cost about four times the average per capita personal disposable income. By 2006, the median house cost about seven times per capita personal disposable income. As of June 2010, this ratio declined to about 4.4, not much higher than the 4.0 average recorded during the 1990s (Fig. 2).

This ratio, however, only measures current house prices against current per capita personal disposable income. Thus, it does not capture the total liability of all the mortgages held by the consumer sector. Comparing this stock of total mortgage liability against per capita personal disposable income gives a more complete picture of the consumer sector's total debt burden of house ownership. As Fig. 3 indicates, this total liability decreased only slightly from the recent peak levels. Here lies the fundamental imbalance: many of the homeowners who purchased their homes during the 2003–2006 period are carrying mortgages that require a significant portion of their income.

[1] Standard & Poor's (2010), S&P Case-Shiller Home Price Index. U.S. Department of Commerce, Bureau of Economic Analysis (2010), National Income and Product Accounts of the United States, Table 2.1. U.S. Census Bureau (2010): New Residential Sales.

There are several data sources for U.S. house prices. *The S&P Case-Shiller Home Price Index* is used here as it represents the broadest measure. Another broadly used index of house prices is the U.S. Census Bureau: New Residential Sales. The Census index shows a more moderate increase in house prices.

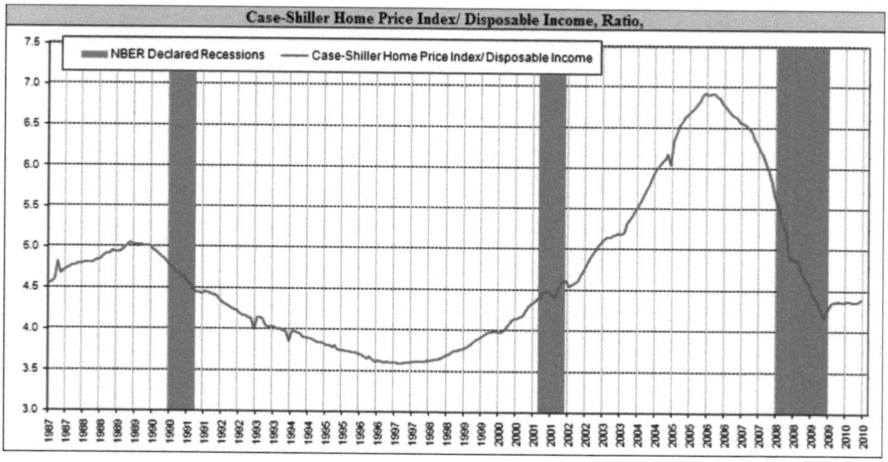

Fig. 2 House prices to per capita personal disposable income (Source: Standard & Poor's, (2010) *S&P Case-Shiller Home Price Index*, U.S. Department of Commerce, Bureau of Economic Analysis, (2010) *NIPA*, and SBA calculations)

3 The Spreading of the Crisis

The U.S. financial sector became more concentrated in mortgages and, during the 1990s, mortgage assets accounted for about 25% of commercial banks' total assets. By 2007 this ratio peaked at 32% (see Fig. 4).

Mortgage assets held by commercial banks, however, do not fully capture the financial sector's total exposure to mortgages. Financial institutions often convert these mortgages into *Mortgage Backed Securities (MBS)*. Portions of these are commonly sold to other institutions, while other portions are retained as security assets. By converting mortgages into securities, financial institutions are able to take advantage of capital requirements that give MBS preferential treatment, increasing their capital base and profit margins. In addition, yields from these MBS are classified as *Non Interest Income,* even though they originated from the interest income of the mortgages that backed the MBS. Investments in MBS made it difficult to ascertain the banking sector's true concentration to mortgage assets.[2] The financial sector expanded and

[2] For example, to ascertain a bank's concentration to *mortgage assets* one now had to dig deeper in regulatory filings, and were possible, separate MBS from other types of securities. While this, rarely done, investigation may have revealed the bank's concentration to MBS, it would not have revealed the bank's total concentration to *Interest Income*. This later differentiation was no longer obtainable from banks' regulatory findings, but could only be obtained via a detailed bank audit. As a result, investors' bank valuations, that over the past 20 years had slowly moved to income-based rather than asset-based valuation, were often in the dark when it came to understanding the banking sector's full exposure to mortgages.

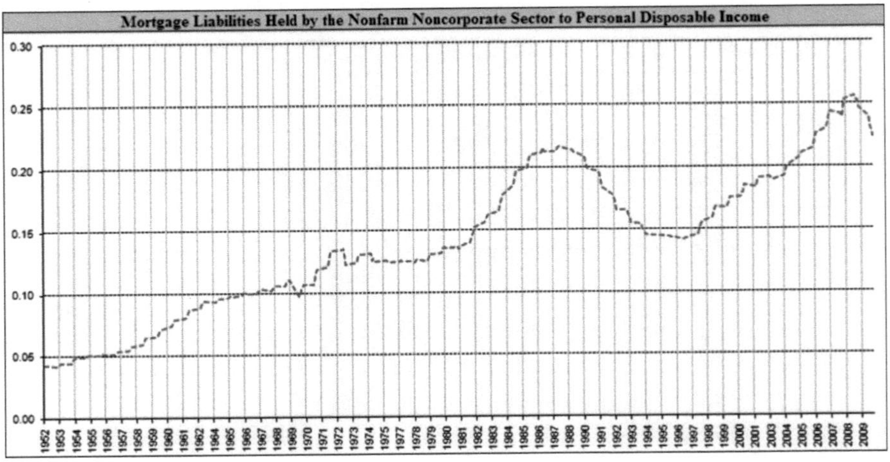

Fig. 3 Nonfarm noncorporate sector's Mortgage liabilities to per capita personal disposable income (the noncorporate sector includes the consumer and non-profit sectors) (Source: Board of Governors of the Federal Reserve System (2010f), *Flow of Funds Accounts of the United States*, U.S. Department of Commerce, Bureau of Economic Analysis (2010), NIPA, and SBA calculations)

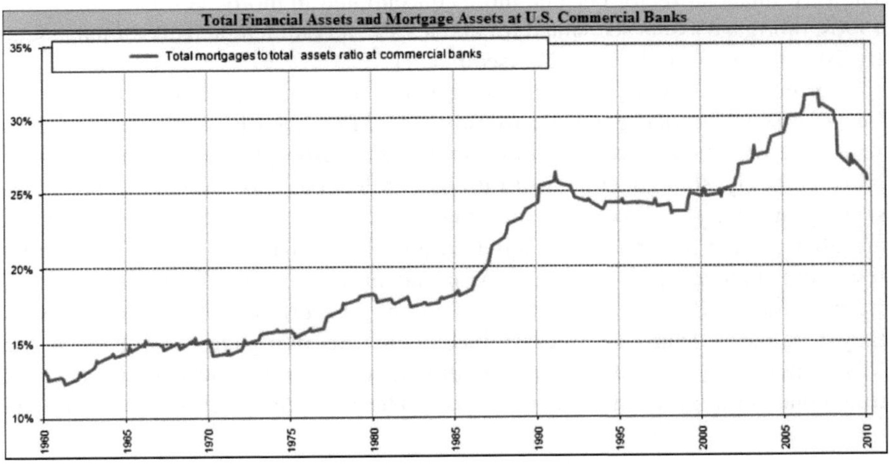

Fig. 4 Total financial assets and Mortgage assets at U.S. Commercial banks (Source: Board of Governors of the Federal Reserve System (2010f), *Flow of Funds Accounts of the United States* and SBA calculations)

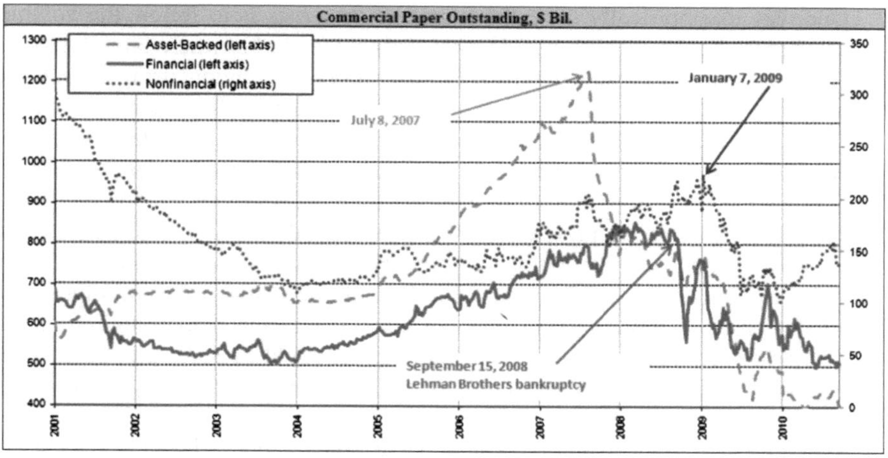

Fig. 5 Commercial paper outstanding, billions of dollars (Source: Board of Governors of the Federal Reserve System (2010e), *Commercial Paper*)

experienced higher profit margins, but it's asset base was opaque, making it difficult to ascertain risk concentration.

4 From an Asset-Based Crisis to an Institutional-Based Crisis

Early in 2007, the market began to question the riskiest type of mortgages, *subprime mortgages*. These mortgages were perceived as riskier given that their looser underwriting standards, among other things, allowed a higher total debt to income ratio. The proponents of subprime lending rationalized this underwriting by arguing that the borrower's debt leverage was but one portion of a mortgages' total risk calculus, as the borrower could easily and quickly sell the house at a profit. In essence, homes were treated as near liquid assets, even in the short-term. This did not seem to be a problem as long as housing prices were rising, as indeed they had been doing for nearly 10 years By January 2007, housing prices had come off the dizzying heights of the previous several years and turned negative, placing this rationale under closer scrutiny.

As the year unfolded, however, a clear picture of these assets was not forthcoming. Financial analysts were not able to provide a speedy or clear assessment of MBS exposure to subprime mortgages, placing downward pressure on MBS prices. This lack of clarity quickly spread to other asset classes, including most mortgages and consumer loans such as automobile, student, and credit card loans. As a result, the downward price-pressure spread to all *Asset Backed Securities (ABS)*.

One way to empirically trace the spread of the crisis is to follow the volume changes in the *Commercial Paper Market*. This is the market where large financial and nonfinancial firms obtain a portion of their short-term funding using some form of collateral (see Fig. 5). The importance of this and other wholesale capital markets had been growing steadily over the previous 20 years, as some financial institutions began to fund a greater portion of their capital needs from these less expensive wholesales markets rather than from more costly retail deposits.[3]

This shift in funding preferences was the critical linchpin that encouraged other fundamental transformations in financial products and institutions. The capacity to bundle and sell a group of assets, allowed financial institutions that originated credit to shorten the maturity of these assets, thus converting long-term illiquid assets into short-term liquid securities that could be sold in bulk to other financial institutions, that in turn could resell them on the secondary market like any other securities.[4]

This product-transformation capacity gave lenders two advantages. First, it allowed them to accelerate the turnover of their capital base by allowing them to quickly convert and re-convert a given capital base from cash into credit-assets and back to cash again, thus capturing servicing fees at a faster pace. It also allowed lenders to sell their ABS holdings and substitute their higher-cost retail deposits with a lower-cost funding from the commercial paper market.

This alternate source of funding provided the incentive for financial institutions from both ends of the spectrum to move to the lightly regulated "Shadow Banking Sector."[5] Large investment banks with high wholesale borrowing capacity but little to no retail capacity, purchased non-depository lending institutions. Commercial banks, with large retail lending capacity, began to increasingly fund their capital needs from less expensive wholesale capital markets.

These two fundamental transformations in the financial markets resulted in two further changes, one of which was unforeseen. First, credit origination risk was shifted from the originators to the rest of the financial sector, including mutual funds, money-market funds, pension funds, etc. Second, the financial sector

[3] For example, during the 2000–2009 period, interest rates on 1-Month Certificate of Deposits, one of the most common forms of retail savings deposits, was 0.09% higher than the interest rate paid by financial institutions to obtain 1-month funds in the commercial paper market. In addition, to this slight market-cost advantage, financial institutions purchasing their capital via the commercial paper market gained a more important operational-cost advantage by gathering funds without the necessity of maintaining an expensive retail branch depository structure. These two cost advantages were the primary financial incentives for the development of the "shadow banking" system. For data on these and other interest rates see Board of Governors of the Federal Reserve System (2010i), *Selected Interest Rates*.

[4] This process of bundling a group of assets and selling different tranches of the bundle as securities became a common practice to transform many types of credit originations such as student loans, auto loans and credit card loans.

[5] For a more detailed discussion on the development of the Shadow Banking System see Date and Konczal (2010).

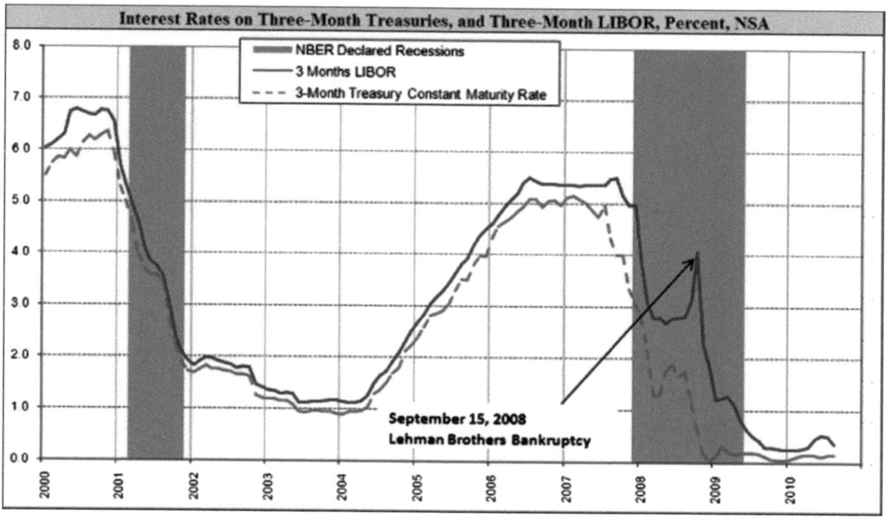

Fig. 6 Interest rates on 3 months U.S. Treasuries and LIBOR (Source: Board of Governors of the Federal Reserve System (2010i), *Selected Interest Rates*, Moody's Economy.com (2010))

became increasingly funded by transient wholesale capital markets rather than more stable retail deposits. Hence, credit originators' funding became ever more dependent on the Fund Industry which, in turn, relied heavily on the opinions of *Rating Agencies*. When rating agencies were not able to quickly provide a clear assessment of MBS and ABS risk, fund managers sold these assets in mass as shown in Fig. 5. The volume of *Asset Backed Commercial Paper Outstanding*, the dashed line in Fig. 5, peaked in August 2007, and continued to drop precipitously thereafter.

Hence, the financial sector experienced a pronounced lack of confidence in mortgages and ABS. This notwithstanding, investors were still willing to lend to financial institutions as long as the collateral was other than ABS. Thus, the volume of non-ABS commercial paper issued by financial institutions, the solid line in Fig. 5, remained at peak levels up to September 2008. With the Lehman Brothers bankruptcy, however, investors became reluctant to accept even these types of collateral. Thus, this market contracted as well. The crisis had now shifted from one based on a specific type of financial asset to one that questioned the liquidity and possibly the solvency of some of the largest U.S. financial institutions.

By January 2009, after a year of economic contraction, the financial crisis had spread further and affected the rest of the economy. At this point investors' pessimism extended further into the real economy and reduced their exposure to commercial paper issued by *Nonfinancial Institutions* as well, the dotted line in Fig. 5. The crisis now represented a near systematic freezing of all major sectors of the U.S. credit markets.

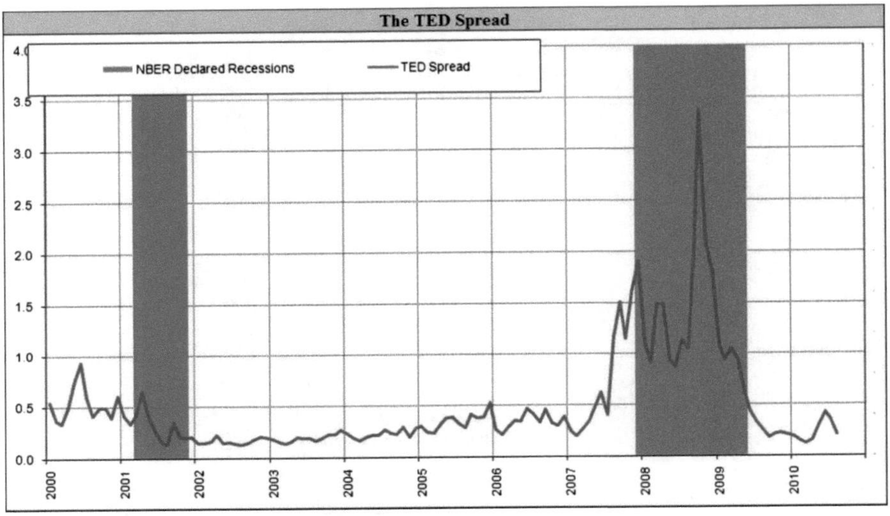

Fig. 7 The TED Spread (Source: Board of Governors of the Federal Reserve System (2010i), *Selected Interest Rates*, Moody's Economy.com (2010), and SBA calculations)

As the volume in the short-term financial markets contracted, interest rates became ever more volatile. Thus, as the Federal Reserve reduced its regulatory short-term rates to respond to the widening crisis, other short term rates came down as well, but not always at the same pace. Figure 6 shows the interest rate for *3-Month U.S. Treasuries*, the dashed red line, considered the safest investment in the U.S. financial markets, and the *3-Month London Interbank Offered Rate* or *LIBOR*, the solid green line. The LIBOR is the rate large financial institutions commonly pay to obtain short-term credit. The space between these two lines can be considered the premium above the rate for Treasuries of similar maturities that financial institutions pay to obtain short-term credit on the short-term inter-bank markets. This spread is commonly referred to as the *TED Spread* and is shown in Fig. 7.[6]

As is evident from Fig. 6, LIBOR rates did not always move in line with changes in safer short-term assets, like U.S. Treasuries, but at times came down more slowly, and during the collapse of the investment firm Lehman Brothers, LIBOR rates actually rose. As a result, the "risk" premium financial institutions had to pay to obtain short-term credit rose from the traditional ½% to a historically high 3.4% (see Fig. 7). This was a steep price for financial institutions, and as a result they responded by reducing credit to the rest of the economy.

[6]TED is the acronym formed by the U.S. Treasury and the trading symbol, ED, for Eurodollar Futures Contracts.

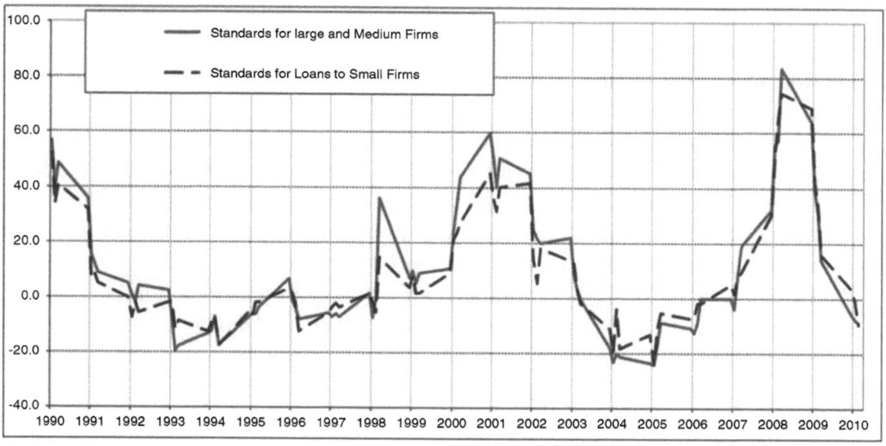

Fig. 8 Senior loan officer opinion survey on bank lending practices (Source: Board of Governors of the Federal Reserve System (2010j), *Senior Loan Officer Opinion Survey on Bank Lending Practices*)

5 The Crisis and Its Impact on U.S. Small Business Financing

A detailed analysis of the changes in demand and supply for small business credit is hampered by a meaningful lack of data. While there is ample quarterly and even monthly data on the supply of consumer loans, high frequency data on the supply and demand for business loans is much harder to obtain. This is especially true for high frequency data on privately held small businesses.[7] This lack of data has significantly impeded our understanding the recent credit contraction has had on small business credit markets.[8] Notwithstanding these data limitations, it became clear that credit to small businesses was being reduced at a pace significantly faster than the contraction of credit for large firms.

The Federal Reserve publishes the *Senior Loan Officer Opinion Survey on Bank Lending Practices*, a quarterly survey on lending standards for approximately 60 of the largest financial institutions under its jurisdiction.[9] Some of the survey's results are displayed in Fig. 8. Interpreting these lending standards can be a bit

[7] For example, the Federal Financial Institutions Examination Council (FFIEC), as part of the *Community Reinvestment Act of 1977*, once a year publishes loan-origination data, including business loans, reported by various financial institutions. For further detail see Federal Financial Institutions Examination Council (2010), http://www.ffiec.gov/

[8] For a detailed discussion on the lack of data on small business credit markets and its impact on the policy making process, see (Congressional Oversight Panel, 2010).

[9] See Board of Governors of the Federal Reserve System (2010j), *Senior Loan Officer Opinion Survey on Bank Lending Practices*.

complicated. Generally speaking, as standards move above the zero line, one can infer that these financial institutions are extending *less* credit to businesses. As standards move below the zero line, one can infer that these institutions are *increasing* credit to businesses.

A close analysis of this dataset indicates that during this cycle, financial institutions tightened credit to small firms, for five quarters longer than credit to large firms. This pattern is the opposite of one set during the 2000–2001 credit tightening cycle, when financial institutions tightening focused on large firms. This was the first indication that this cycle would be hardest on small firms.

Unfortunately, these lending standards are un-weighted diffusion indices, and as such do not provide any information on the level of or change in the *volume* of lending supplied by these institutions. At best we can infer a general directional change in the level of credit supply.

A rough sense of the lending level supplied to small firms by depository institutions, which provide about 64% of traditional credit to small businesses, can be obtained from the Federal Deposit Insurance Corporation (FDIC), Call Reports.[10] The FDIC insures the deposits held by the institutions under its charter. Its Call Reports provide end-of-quarter loan balances held by these institutions. This dataset shows that as of the third quarter of 2010, depository institutions holdings of small loans declined to $ 630 billion, representing an 11% drop from the peak level of $ 711 billion held at the end of the second quarter of 2008.

This dataset most likely understates the decline in small business lending by these institutions. *End-of-quarter balances* of small loans are not the supply of small loans provided *during the quarter*. In essence, these are inventory data that are often erroneously referred to as supply data. Because financial institutions can sell the credit they originate during the quarter, end-of-quarter inventory is a poor indicator of supply provided during the quarter. Thus, during the pre-crisis period, when the secondary markets were at their peak, financial institutions could easily sell the loans they produced during the quarter, thus showing little to no change in their balance positions at the end of the quarter (Fig. 9).[11]

The data availability on small business credit *demand* is just as, if not more limited. Whatever data is available, do indicate that demand did contract significantly during the recession (see Figs. 10 and 11). The two major data sources on small business credit demand, however, portray a contradictory picture as to

---

[10] For data on balances held by depository institutions see Federal Deposit Insurance Corporation (2010), *Call Reports*. For data and discussion on the share of credit supplied to small businesses see U.S. Small Business Administration (2009b) *Small Business in Focus: Finance*.

[11] For example, an institution may underwrite $100 million in small business loans in one quarter, and sell 50 million of those loans on the secondary market. As a result, the institution's balance sheet would increase by $50 million. In the following quarter, the same institution may underwrite only $75 million in small business loans, and sell only 25 million of these loans on the secondary market. Once again, the institution's balance sheet would increase by another 50 million. Hence, the balance sheet position continues to expand at the same $50 million pace, when in fact there was a $25 million reduction in its underwriting.

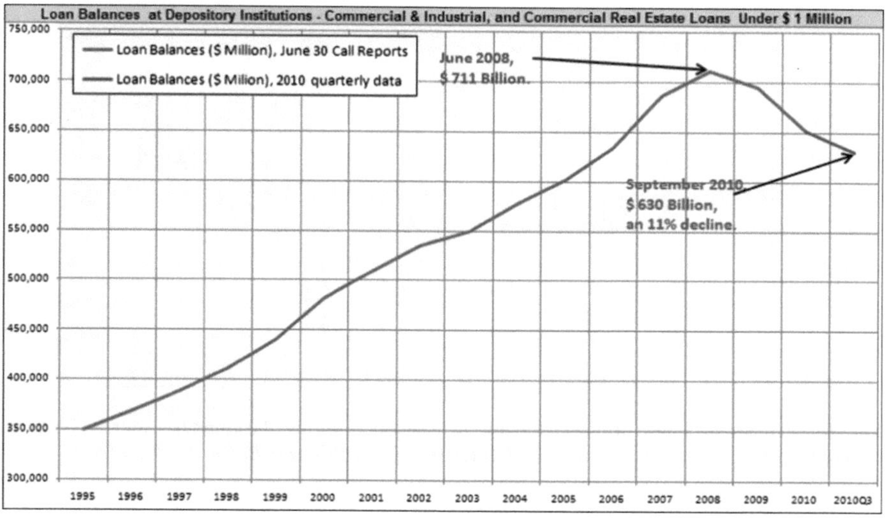

Fig. 9 Loan balances at depository institutions – commercial & industrial, and commercial real estate loans under one million dollars (Source: Federal Deposit Insurance Corporation (2010), *Call Reports*)

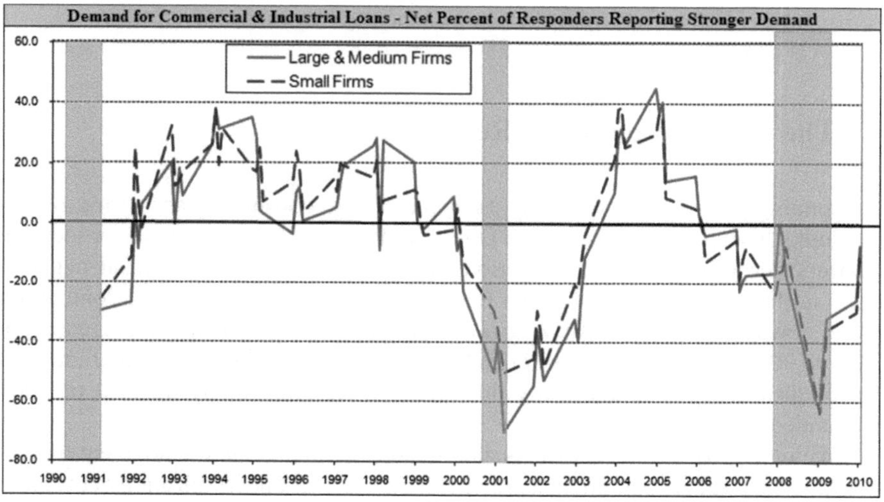

Fig. 10 Demand for commercial and industrial loans by firm size (Source: Board of Governors of the Federal Reserve System (2010j), *Senior Loan Officer Opinion Survey on Bank Lending Practices*)

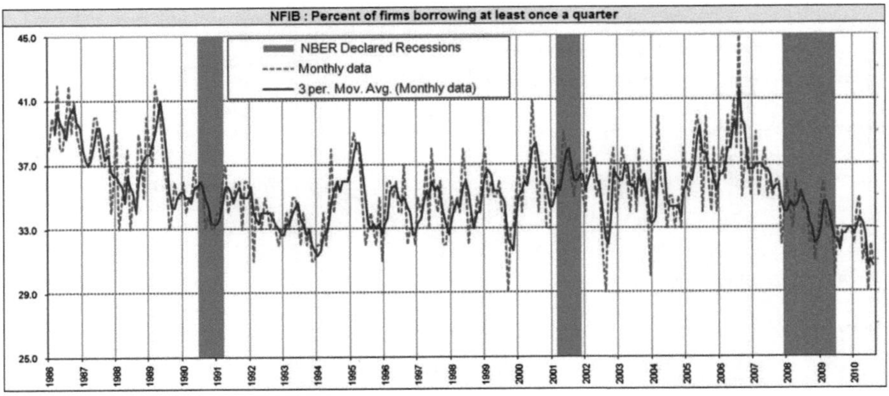

Fig. 11 Percent of small firms borrowing at least once during the quarter (Source: National Federation of Independent Business (NFIB) (2010), *Small Business Economic Trends*)

whether small business credit demand rebounded after the recession ended in July 2009. The Federal Reserve data indicate that credit demand by small, as well as large, firms followed a typical cyclical contraction-pattern and rebounded after the recession ended. On the other hand, data from the National Federation of Independent Business (NFIB) indicate that small business credit demand continued to decline after the recession ended.

6 The Contraction in the Real Economy

Responding to the ever-widening financial crisis, consumers and firms turned cautious, driving down Real Gross Domestic Product (GDP) for four consecutive quarters during the 2008–2009 period. Real GDP reached a bottom at negative 6.8% in the fourth quarter of 2008. By the end of 2009, the economic contraction drove down the *Labor Market Utilization Rate,* which historically averaged about 94%, to 90%, and seemed stuck at this level.[12] The Utilization Rate of the Industrial Sector also dropped dramatically during the recession, but by the end of 2009 rebounded strongly (see Fig. 12).

Like other post 1980s recessions, production rebounded but employment utilization languished at the bottom for a prolonged period. This underutilization of the labor market would become one of the federal government's main challenges in this crisis.

[12] The Labor Market Utilization Rate is close to the inverse of the Unemployment Rate. So, as this ratio increases the unemployment rate tends to decline.

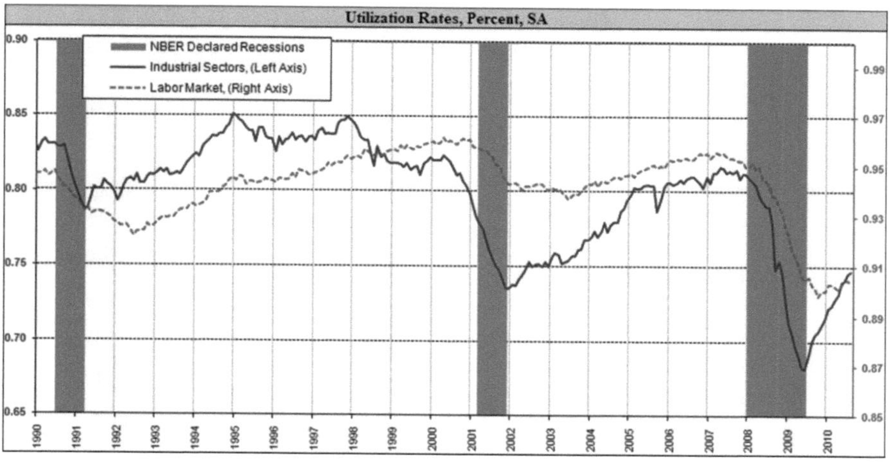

Fig. 12 Utilization rates in the industrial sectors and labor market (Source: Board of Governors of the Federal Reserve System (2010g), *Industrial Production*, U.S. Bureau of Labor Statistics (BLS), *Current Employment Statistics*, U.S. Bureau of Labor Statistics *Current Population Survey*, and SBA calculations (U.S. Bureau of Labor Statistics (2010b, c)))

7 Conventional and Non-conventional Federal Government Responses

The conventional first response to financial instability and economic contraction in the U.S. has been to adjust *Regulatory Short-Term Interest Rates* such as the *Federal Funds Rate and Federal Reserve Discount Rate.*[13] In this regard, this crisis has been no different. Hence, by July 2007 the Federal Reserve started to aggressively reduce the Effective Federal Funds Rate and progressively reduced it each month so that by January 2000 it reached a near zero level of 0.15%. The Federal Funds Rate has remained at this extreme low level ever since. Never before has the Federal Reserve maintained such an accommodative interest rate policy for so long.

These unprecedented aggressive interest rate maneuvers, however, were not enough. There were still two fundamental challenges that needed to be quickly and forcefully addressed. The first immediate task was to unfreeze the financial markets by

[13] The *Federal Funds Rate* is the interest rate a qualified financial institution pays to borrow short-term funds from another financial institution with excess reserves at the Federal Reserve. The Federal Reserve manipulates this rate by purchasing or selling short-term Treasuries held by financial institutions. The *Discount Rate* is the rate the Federal Reserve charges financial institutions to borrow short-term funds from its own cash-reserves. The Federal Reserve unilaterally sets the Discount Rate. For further details see the Board of Governors of the Federal Reserve System (2010h), *Monetary Policy*.

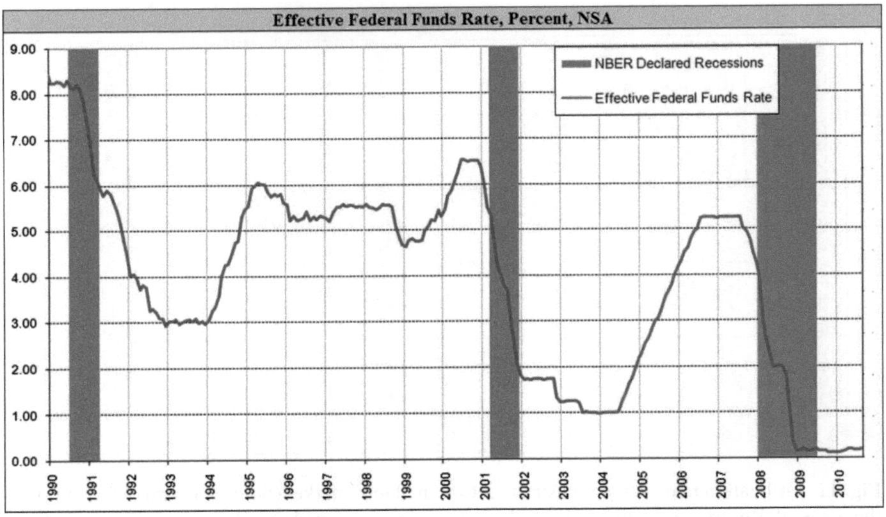

Fig. 13 Effective federal funds rate, percent (Source: Board of Governors of the Federal Reserve System (2010i), *Selected Interest Rates*)

Federal Government Response to the Crisis	Number of Actions	Originally Committed, $ bil.	Currently Provided, $ bil	Ultimate Cost, $ bil.
Federal Reserve	22	6,700	1,798	15
Treasury	2	560	345	305
FDIC	6	2,913	328	75
Federal Housing Administration	2	100	150	26
Congress	6	1,668	892	1,169
Total	38	11,941	3,513	1,590

Fig. 14 Federal government responses to the crisis

cajoling volume back to more normal levels. The second task was to reduce risk premiums to more customary ranges. The third task was to stimulate the real economy.

The federal government responded to these three challenges with numerous enormous unconventional measures. The volume and price challenges in the financial sector were addressed by injecting massive amounts of liquidity and price assurances into the financial sector, while the economic contraction was addressed by injecting an enormous amount of government spending into the economy.

In total, the federal government undertook 38 major actions. Thirty-two actions were intended to ensure sufficient liquidity in the financial markets and asset price stability in the financial markets and real sector of the economy (see Fig. 13). The remaining six major actions were intended to stimulate the real sectors in the economy.[14]

In all, the federal government committed approximately $12 trillion, or 86% of GDP, to these efforts. Some experts, however, estimate the final costs may be reduced to be around $1.6 trillion, or 11% of GDP. The U.S. GDP during 2008–2009 was about $14 trillion (Fig. 14).

8 Non-conventional Responses by the Federal Reserve

The Federal Reserve was one of the lead agencies to address these economic challenges. In addition to lowering its regulatory short-term interest rates to the unprecedented near-zero level, it undertook several non-conventional actions to stabilize the various financial markets. In so doing, it significantly extended its operations and influence well beyond the large commercial banks and bank holding companies traditionally under its jurisdiction. The Federal Reserve now provided support to, and as a result now shared oversight over, investment banks, brokerage institutions, and some large insurance companies.[15] These non-conventional interventions were intended to accomplish four basic objectives.

The first objective was to foster liquidity and price stability of domestic bonds and equities. This objective was tackled with the introduction of several programs during the first half of 2008. Among these were the *Term Securities Lending Facility*, *Primary Dealer Credit Facility*, and the *Term Securities Lending Facility Options Program.*[16] Under these programs the Federal Reserve provided low risk

[14] The data for Fig 14 and this section was primarily derived from the following sources: (Blinder and Zandi 2010), (The Federal Reserve Bank of St. Louis 2010), (Guillén 2009), (Congressional Oversight 2010), (The White House, Council of Economic Advisors 2010).

[15] As a result of the *Glass-Steagall Act of 1933*, the financial sector was separated into two distinct sectors. The *Intermediary Sector* was to provide safe deposit and reliable loans, and was to include commercial banks, depository and savings institutions, and non-bank banks. The *Direct Capital Market* was to provide riskier investment products, and was to include institutions such as investment banks, brokerage houses, and some activities of insurance companies. While the post 1980 financial deregulation acts (the *Depository Institutions Deregulation and Monetary Control Act of 1980, The Garn-St Germain Depository Institutions Act of 1982* and the *Gramm–Leach–Bliley Act of 1999*) largely dissolved the functional distinctions of these institutions, the fragmented regulatory infrastructure remained.

[16] The *Term Securities Lending Facility (TSLF)* was a facility designed to lend up to $200 billion of Treasury securities on a weekly basis for 28-day periods to primary dealers in exchange for less liquid collateral. The *Primary Dealer Credit Facility* was an overnight loan facility that provides funding to the New York Fed's Primary Dealers by accepting collateral beyond the usual investment-grade debt securities. The *Term Securities Lending Facility Options Program*

U.S. Treasuries to Registered Broker Dealers in return for collateral beyond the usual investment-grade debt securities. In so doing, the Federal Reserve took out of circulation a certain portion of financial assets, which the market now considered too risky or unpredictable, and at the same time, it bolstered the broker dealers' capital base with Tier One Capital, i.e. Treasures. The same "swap" principle of cash or low-risk assets for high-risk assets was used for the remaining objectives.

The second objective was to shore up large, systematically important financial institutions. This objective was tackled in a cooperative effort with the FDIC, and the U.S. Department of Treasury whereby these federal agencies injected either cash or Treasuries in these institutions in exchange for equities and options.[17]

'The third objective was to enhance the liquidity and price stability of Asset Backed Securities traded in the Commercial Paper Market. Again the same principal was used where federal agencies swapped questionable assets for either cash or treasuries.[18]

The final objective was to ensure that the global currency markets had sufficient quantities of U.S. dollars. This objective was achieved by establishing Reciprocal Currency Agreements with various other central banks, whereby the Federal Reserve provided U.S. dollars to other central banks.[19]

The fast pace of events and the volatility in the financial markets placed the U.S. Department of Treasury in the critical role not only by being a significant player in determining the specific nature of these unprecedented interventions, but also by quickly transmitting very large sums of funds to other agencies such as the Federal Reserve, the FDIC, Federal Housing Authority (FHA), and the SBA.

These and other major agencies used their additional funding authorities to increase the operations in their respective domains. The FHA, for example, deployed additional funding for mortgage guarantees, and initiated mortgage foreclosure prevention programs. The FDIC increased the limit of insured deposit accounts from $100,000 to $250,000. The FDIC also played a critical role in the various bailouts of financial institutions.

While these agencies proposed these new unconventional interventions, it was the job of the U.S. Congress to enact legislation that would provide the authority and funding for these maneuvers. Among the numerous laws enacted during the

significantly expanded the TSLF. For further details see the following documents from the Board of Governors of the Federal Reserve system: March 11, 2008 Press Release, March 18 Press Release, and the *Term Securities Lending Facility (TSLF) and TSLF Options Program (TOP)* (Board of Governors of the Federal Reserve System 2010a, b, c).

[17] These maneuvers are commonly referred to as bailouts.

[18] Among these maneuvers were the *Asset Backed Commercial Paper Money Market Mutual Fund Liquidity Facility*, *Commercial Mortgage Backed Securities Program*, *The Commercial Paper Funding Facility*. For further details see (Blinder and Zandi 2010) and (The Federal Reserve Bank of St. Louis 2010).

[19] Ibid.

2008–2010 period to address the financial and economic crisis, four stand out in importance and scale.[20]

The first of these laws was the *Economic Stimulus Act of 2008*. This law was enacted early in the crisis, on February 7, 2008, when it became clear that personal income was insufficient to support many of the mortgage and other consumer debt obligations. It thus provided $152 billion that extended direct tax rebates to low and middle-income individuals, who were expected to quickly spend the bulk of this additional cash. In addition, it accelerated capital depreciation to small businesses, which for the past 15 years had created some 65% of net new jobs.

This fiscal maneuver, along with the rapid reduction of interest rates orchestrated by the Federal Reserve had an immediate, but temporary impact. The annual growth rate of nominal personal disposable income did accelerate from a long-term average of about five to a peak of 9% in May 2008, and the change in GDP did turn positive for the first quarter of 2008.

These maneuvers were not enough, however, and the economy entered into a severe contraction. By the third quarter of 2008 the financial markets were quickly seizing, and on October 3rd Congress turned its attention to stabilizing the financial sector by enacting the *Emergency Economic Stabilization Act of 2008*. The primary purpose of this act was to provide $700 billion to fund the *Troubled Asset Relief Program* (TARP). The primary function of the TARP was to purchase or insure troubled assets so as to avert the potential failure of large financial institutions. Troubled assets were broadly defined as any financial instrument the purchase of which is necessary to promote financial market stability. Sellers of troubled assets would be required to issue warrants to the Treasury exercisable for nonvoting shares. Participating institutions would also be subject to the Treasury's new standards for executive compensation and corporate governance.

By the fourth quarter of 2008, real GDP had contracted sharply for four consecutive quarters, and Congress, on February 17, 2009, enacted the *American Recovery and Reinvestment Act of 2009*. The Act attempted to stimulate the economy by providing $787 billion for tax incentives, social programs, infrastructure investment, energy efficiencies and investment programs, and small business related programs.

By the middle of 2010, the economy had begun to recover ever so slowly, but credit to small businesses was still severely constrained. Responding to this need, Congress enacted the *Small Business Jobs and Credit Act of 2010*. The Act established a $30 billion *Small Business Lending Fund*, provided $12 billion in tax incentives to small businesses, and granted additional funding for the SBA.

[20] For links to these legislations see (Federal Reserve Bank of St. Louis 2010). For further discussions on these legislations see (Blinder and Zandi 2010) and the (White House, Council of Economic Advisors 2010).

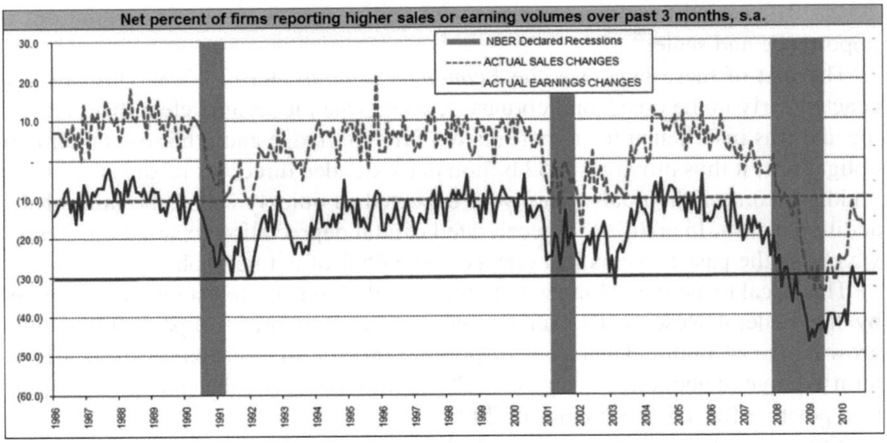

Fig. 15 Sales and earnings at small firms (Source: National Federation of Independent Business (NFIB) (2010), *Small Business Economic Trends*)

9 The Financial and Economic Crisis of 2008–2009 and Its Impact on U.S. Small Businesses

The SBA broadly classifies small businesses as any firm with 500 or fewer employees.[21] These firms account for more than 26 million businesses, or 99% of all firms. They employ slightly over half of the private sector's employees, pay about 44% of the total private sector payroll, generate about 65% of net new private sector jobs, and create more than half of the nonfarm private Gross Domestic Product.[22]

The impact of the economic contraction quickly transmitted to the sales and earnings of small firms. Once again, the lack of current data makes it nearly impossible to determine the actual recent decline in sales and earnings of privately held small firms. Fortunately, the NFIB does provide un-weighted diffusion indices for sales and earnings at small firms (see Fig. 15).[23] This dataset indicates that small

[21] The SBA has two different approaches for defining small firms. The first approach is to define any small firm with less than 500 employees as "small." This practice was first established by the *Small Business Act of 1953*. However, the same Act required the SBA to establish a size standard that "should vary to account for differences among industries." Second, the Act called on the SBA to "assist small businesses as a means of encouraging and strengthening their competitiveness in the economy." These two considerations are the basis for the SBA current methodology for establishing small business size standards. For further details see The U.S. Small Business Administration (2009a) *SBA Size Standard Methodology*.

[22] For more details on the importance of small businesses in the U.S. economy see The U.S. Small Business Administration (2010a), *Frequently Asked Questions*.

[23] The U.S. Internal Revenue Service does provide some financial data by firm asset size at an annual frequency with several years lag. Thus we will eventually be able to get a more detailed

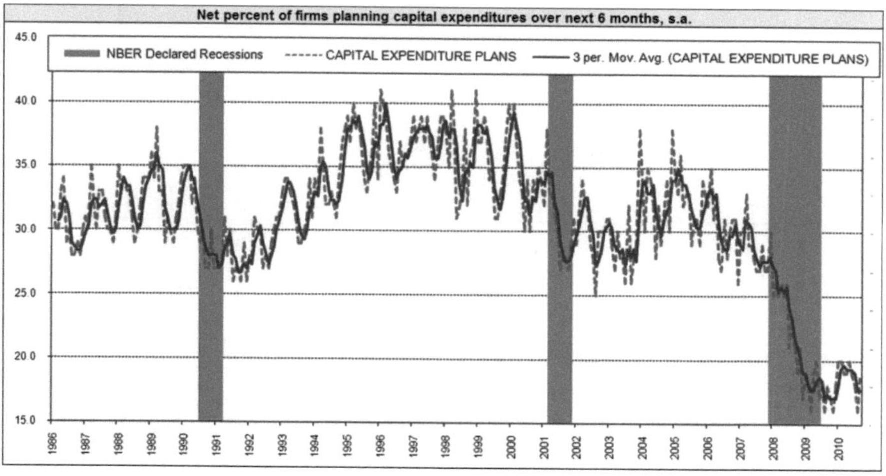

Fig. 16 Small firms' capital expenditure plans (Source: National Federation of Independent Business (NFIB) (2010), *Small Business Economic Trends*)

firms reported that both actual sales and earnings reached historical lows during the crisis, and only after some 13 months of improvements did these indicators reach the lows of previous recessions. By contrast, the profit margins of the corporate sector as a whole rebounded strongly. For example, After Tax Net Corporate Earnings as a percent of GDP reached a cyclical low of 5.3% during the fourth quarter of 2008, but by the second quarter of 2009 had rebounded to 9.6%, well above the historical average of 6.0%.[24]

The stark difference in performance when viewed by firm-size is a dominant feature of this crisis. Thus, large firms that had exposure to faster-growing international markets and possibly to the primary commodity sectors, which experienced noticeable price appreciation, were able to record strong profit gains. U.S. small firms, that were heavily dependent on the slower growing domestic economy, and to the retrenching U.S. consumer sector, experienced only modest recovery from their historical lows.

The pronounced decline of small firms' sales and earnings had a devastating impact on their capital expenditure plans, which remained at the historical lows reached during the second half of 2009 (see Fig. 16). Likewise small firms reduced their labor force at a faster rate than large firms.

The U.S. Bureau of Labor Statistics publishes employment by firm size on a quarterly basis, but with a three quarter lag. Hence, the fourth quarter of 2009 is the

---

empirical sense of the impact of the 2008–2009 recession on small businesses. For annual data by firm asset size see The U.S. Internal Revenue Service (2010), *Statistics On Income*.

[24] See the U.S. Department of Commerce, Bureau of Economic Analysis (2010), *National Income and Product Accounts of the United States*.

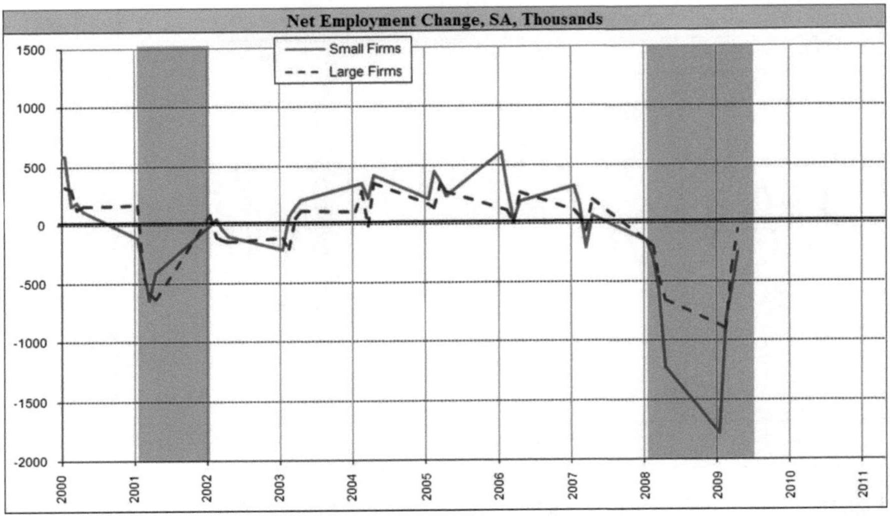

Fig. 17 Net employment change by firm size (Source: U.S. Bureau of Labor Statistics (2010a), *Business Employment Dynamics*)

most current available data (see Fig. 17). A detailed analysis of this dataset, shows that very small firms, those with fewer than 10 employees started contracting net employment during the second quarter of 2007. Thus, by the fourth quarter of 2009, firms with fewer than 500 employees accounted for the loss of 5.7 million net job losses, or 62% of the total net job contraction. Large firms, on the other hand, account for 3.5 million net job losses or 38% of the total net job losses.

By contrast, during the employment decline of the 2001–2003 period, small firms accounted for 45% of net job losses, and large firms accounted for 55% of the total net employment losses. In this respect, this recession is more like the employment contraction of the 1990–1992 period, when small firms accounted for 80% of the net job losses and large firms accounted for the remaining 20%.[25]

10 The U.S. Small Business Administration Role

Given the pronounced decline in sales, earnings and employment at small firms, the primary objectives of the federal government with respect to small businesses were first to increase the supply of credit to small firms, and second to fortify the balance sheet of small firms so as to increase their capacity to borrow, which would in turn

[25] Data from the private firm Automatic Data Processing Inc. (ADP), show that as of early 2010 small firms had started to gain net new jobs. See ADP (2010).

would allow them to expand their capital expenditures and employment. The SBA played a critical role in reaching the first objective.

The SBA supports small businesses through three major functions: Capital Access Programs, Government Contracting Programs, and Counseling Programs. This section will focus on the SBA's Capital Access Programs.

The SBA works with approximately 5,000 banks and credit unions, some 250 Community Development Corporations (CDCs), over 170 non-profit financial intermediaries and Community Development Financial Institutions (CDFIs), and approximately 300 small business investment companies (SBICs). These institutions make the decision of whether and at what terms credit is extended to small businesses. They can then request the SBA to guarantee a portion of the credit to be supplied to small businesses. The SBA will grant this request based on a series of lender and borrower criteria. One critical element of these criteria is that the borrower's credit demand must have been rejected by a private sector financial institution.[26] As such, the SBA only guarantees the portion of the small business credit market that is not able to obtain credit, but that does become viable with SBA guarantees.

The SBA Capital Access Program has four major sub-programs that provide guarantees and co-funding for a wide range of products designed to meet the diverse financial needs of small firms throughout their life cycle, starting from small start-ups to established firms.[27]

The largest of these, the *7(a) Loan Program*, provides guarantees for working capital loans up to $5.0 million to new and existing small businesses. The second largest sub-program, the *Certified Development Corporation 504 Loan Program*, provides guarantees and co-funding for loans up to $5.0 million used for the purchases of fixed assets. The SBA *Microloan Program*, provides grant-funding to non-profit intermediaries and CDFI, which in turn offer loans up to $35,000 to small, mostly start-up firms. Finally the SBA's *Small Business Investment Program* provides co-funding and guarantees to SBA's Certified Small Business Investment Companies (SBIC), which in turn leverage SBA contributions with their own funding to provide equity and other long-term investment capital to small businesses (Fig. 18).

The financial and economic crisis had a pronounced impact on the SBA's Capital Access Programs, with the volume of its two largest loan guarantee programs, the 7(a) and 504 programs, declining from a monthly average of $1.7 billion during the 2005–2007 period to a bottom monthly level of $687 million, an approximate 60% decline. The post intervention, average for these two programs rebounded to $1.4 billion.

---

[26] In addition the borrower must be defined as a "small business" under the industry-specific SBA Size Standards, have an ongoing operation, be organized for profit, be located in the United States or U.S. Territories, be 51% owned by a U.S. Citizen or Permanent Resident, and demonstrate a need for credit.

[27] For further details on the SBA's Capital Access Programs see www.sba.gov.

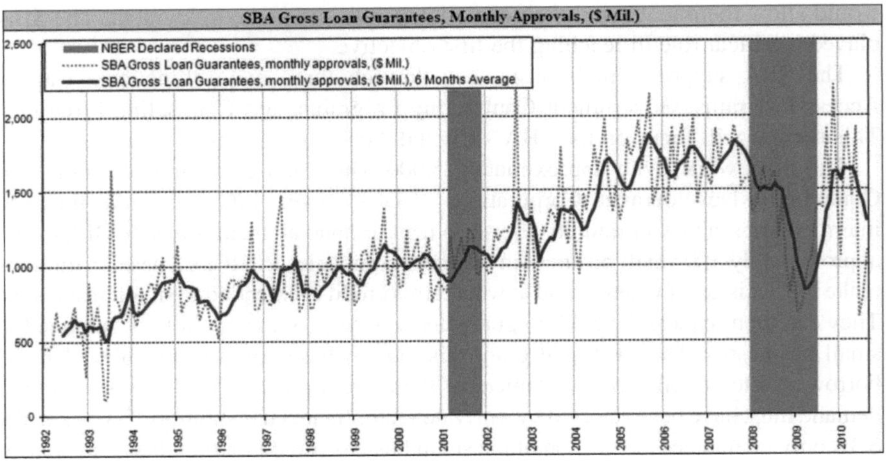

Fig. 18 SBA 7a and 504 loan guarantees (Source: U.S. Small Business Administration (2010b))

To achieve this rebound in volume, the SBA had to address two major challenges. First, it had to provide additional incentives to financial institutions, and second, it had to unfreeze the secondary market for SBA loans.

The SBA addressed the incentive challenge by using the increased funding received from Congress to undertake several program changes. The SBA received an additional $680 million from the *American Recovery and Reinvestment Act of 2009*, nearly doubling its operating budget, and it received an additional $505 million from the *Small Business Act of 2010*.

The Agency used these additional funds to temporarily increase its guarantees from around 75% to 90% of each loan for which SBA provided guarantees.[28] ISBA also temporarily reduced or eliminated the fees it charges financial institutions participating in its loan guarantee programs. The Agency also increased its 7(a) and 504 loan limits from about $2.0 to $5.0 million. The medium sized Express Loan limits were also increased from $350,000 to $1.0 million, and the small Microloan limit was raised from $35,000 dollars to $50,000.

The second problem facing the SBA was that, due to the sharp drop of interbank confidence, the volume in the secondary market for SBA (7a) backed loans dropped sharply just as it did in the commercial paper market. About 40% of 7(a) guaranteed loans are sold in the secondary market.

This market experienced a sharp drop during the second half of 2008, moving from an average monthly level of about $328 million to approximately $100 million during the first month of 2009. In response, on March 16, 2009 the U.S. President

[28] The SBA provides a range of guarantees through its various guarantee products. For a quick reference on the details of its loan guarantee programs see the U.S. Small Business Administration (2010b), *Quick Reference to SBA Loan Guarantee Programs*.

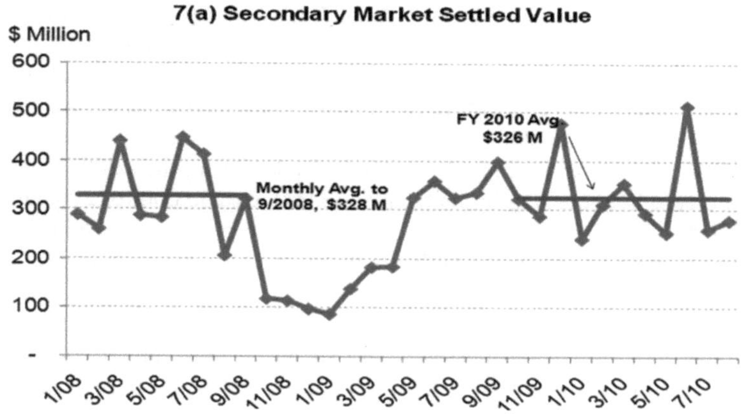

Fig. 19 The secondary market for SBA 7(a) loan guarantees (Source: U.S. Small Business Administration (2010b))

announced that as part of the *Financial Stability Plan* (FSP), the U.S. Department of Treasury would purchase $15 billion of SBA loans on the secondary market.[29]

One of the factors differentiating the FSP and to a certain extent the Term Asset-Backed Securities Loan Facility (TALF), from the TARP, is that the first two programs only purchased "troubled assets" of a recent period. By contrast the TARP primarily purchased "troubled assets" that were originated during the relatively distant 2003–2007 period. Hence, the TARP simply deleveraged financial institutions of their older loans with no preconditions that the deleveraged institutions use their improved capitalization to originate new loans. The FSP and TALF on the other hand "forced the hand" of these institutions to make future loans so as to continue to benefit from the FSP and TALF.

In essence, the government promised to be a buyer of last resort for these recent loans. As a result, investors became quite willing to purchase these assets. Indeed, the guarantee worked so well that the Treasury only intervened in the early part of the program, as the private sector became quite willing to absorb these loans. Thus, by the third quarter of 2010, the average volume for this market returned to the pre-crisis levels (Fig. 19).

The SBA contributed to the fortification of small firms' balance sheets by ensuring that small firms get their fair share of federal government contracts, especially now that federal government contracts would rise steeply as part of the various stimulus programs. The SBA is responsible for ensuring that 23% of federal

[29] The Federal Reserve had already started to purchase some SBA guaranteed loans under the *Term Asset-Backed Securities Loan Facility (TALF)*. For more details see Board of Governors of the Federal Reserve System (2010d), *Term Asset-Backed Securities Loan Facility (TALF) Terms and Conditions*, and The (White House Office of the Press Secretary 2010).

prime contracts go to small businesses. As of August 31, 2010 $10 billion or 32.5% of federal stimulus contract dollars have gone to small businesses.

The numerous tax breaks instituted over the past several years also fortified small firms' balance sheets. The following are some of the major small business related tax changes of the 2008–2010 period: (1) an accelerated first-year depreciation to 50% of the cost of capital purchases made during the 2008–2010 period. (2) An increase in expensing limits from $133,000 to $250,000 for 2008 and 2009, and $500,000 for 2010–2011. (3) A $1,000 new employee tax credit. (5) A tax credit up to 35% of employee healthcare premiums. (6) A tax deduction of up to $10,000 of business start-up costs.

These tax breaks tend to increase after tax net income and cash flow, which the government hopes small businesses would spend on new capital equipment and additional labor employment.

References

ADP (2010) http://www.adpemploymentreport.com/indexsbr.aspx. Accessed 10 Oct 2010

Blinder Alan and Mark Zendi. How the Great Recession Was Brought to an End, July 27, 2010

Board of Governors of the Federal Reserve System (2010a) March 11, 2008 Press Release. http://www.federalreserve.gov/newsevents/press/monetary/20080311a.htm. Accessed 10 Oct 2010

Board of Governors of the Federal Reserve System (2010b) March 18 Press Release http://www.federalreserve.gov/newsevents/press/monetary/20080316a.htm. Accessed 10 Oct 2010

Board of Governors of the Federal Reserve System (2010c) Term securities lending facility (TSLF) and TSLF options program (TOP). http://www.federalreserve.gov/newsevents/reform_tslf.htm. Accessed 10 Oct 2010

Board of Governors of the Federal Reserve System (2010d) Term asset-backed securities loan facility (TALF) terms and conditions. http://www.federalreserve.gov/newsevents/press/monetary/monetary20081125a1.pdf. Accessed 10 Oct 2010

Board of Governors of the Federal Reserve System (2010e). Commercial paper. Accessed 10 Oct 2010

Board of Governors of the Federal Reserve System (2010f). Flow of funds accounts of the United States. Accessed 10 Oct 2010

Board of Governors of the Federal Reserve System (2010g). Industrial production. Accessed 10 Oct 2010

Board of Governors of the Federal Reserve System (2010h). Monetary policy. Accessed 10 Oct 2010

Board of Governors of the Federal Reserve System (2010i). Selected interest rates. Accessed 10 Oct 2010

Board of Governors of the Federal Reserve System (2010j). Senior loan officer opinion survey on bank lending practices. Accessed 10 Oct 2010

Congressional Oversight Panel (2010) The small business credit crunch and the impact of the TARP. http://cybercemetery.unt.edu/archive/cop/20110402035902/http://cop.senate.gov/documents/cop-051310-report.pdf

Date R, Konczal M (2010) Out of the shadows: creating a 21st century glass-steagall. Cambridge Winter Center. http://www.cambridgewinter.org/Cambridge_Winter/Welcome_files/out%20of%20the%20shadows%20030310.pdf. Accessed 10 Oct 2010

Federal Deposit Insurance Corporation (2010) Call reports. www.fdic.gov. Accessed 10 Oct 2010

Federal Financial Institutions Examination Council (FFIEC) (2010) http://www.ffiec.gov/. Accessed 10 Oct 2010
Federal Reserve Bank of St. Louis. The Financial Crisis: A Timeline of Events and Policy Actions, http://timeline.stlouisfed.org/pdf/CrisisTimeline.pdf. Downloaded November 4, 2010
Guillén MF (2009) The global economic & financial crisis: a timeline http://lauder.wharton.upenn.edu/pdf/Chronology%20Economic%20%20Financial%20Crisis.pdf. Accessed 10 Oct 2010
Moody's Economy.com. (2010) http://www.economy.com/freelunch/default.asp?src=economy_quick_links_freelunch. Accessed 10 Oct 2010
National Federation of Independent Business (NFIB) (2010). Small business economic trends. Accessed 10 Oct 2010
Standard & Poor's (2010) S&P Case-shiller home price index. http://www.standardandpoors.com/indices/sp-case-shiller-home-price-indices/en/us/?indexId=spusa-cashpidff–p-us. Accessed on 10 Oct 2010
The White House Office of the Press Secretary (2010) Remarks by the president to small business owners, community lenders and member of congress, March 16, 2009. http://www.whitehouse.gov/the_press_office/Remarks-by-the-President-to-small-business-owners/. Accessed 10 Oct 2010
The White House, Council of Economic Advisors (2010) Economic report of the president. http://www.gpoaccess.gov/eop/2010/2010_erp.pdf. Accessed 10 Oct 2010
U.S. Bureau of Labor Statistics (2010a). Business employment dynamics. Accessed 10 Oct 2010
U.S. Bureau of Labor Statistics (2010b). Current employment statistics. Accessed 10 Oct 2010
U.S. Bureau of Labor Statistics (2010c). Current population survey. Accessed 10 Oct 2010
U.S. Census Bureau (2010) New residential sales. http://www.census.gov/const/www/newressalesindex.html. Accessed on 10 Oct 2010
U.S. Small Business Administration (2009a) SBA Size standard methodology. http://www.sba.gov/content/size-standards-methodology. Accessed 10 Oct 2010
U.S. Small Business Administration (2009b) Small business in focus: finance. http://archive.sba.gov/advo/research/09finfocus.pdf. Accessed 10 Oct 2010
U.S. Small Business Administration (2010a) Frequently asked questions. http://www.sba.gov/advocacy/7495/8420. Accessed 10 Oct 2010
U.S. Small Business Administration (2010b) Quick reference to SBA loan guarantee programs. http://www.sba.gov/sites/default/files/files/LoanChartHQ20110728.pdf. Accessed 10 2010
U.S. Department of Commerce, Bureau of Economic Analysis (2010) National income and product accounts of the United States, Table 2.1. http://www.bea.gov/. Accessed 10 Oct 2010
U.S. Internal Revenue Service (2010) Statistics on income. http://www.irs.gov/taxstats/productsandpubs/article/0,,id¼125133,00.html, Accessed 10 Oct 2010

Access to Finance in the Euro Area: What Are SMEs Telling Us About the Crisis?

Annalisa Ferrando

1 Introduction

Access to finance is widely perceived to be a crucial factor for firms – especially small and medium-sized enterprises (SMEs) – to maintain their day-to-day business as well as to achieve long term growth and investment goals. Hence, experiencing major financing obstacles can be a considerable challenge for enterprises, which in turn can increase credit risks in the corporate sector and also negatively affect productivity in the economy. This seems to be of even more relevance today when, during the current financial crisis, sources of firm financing have become scarcer and the availability of financing instruments has deteriorated.

In this context, it is important to understand whether SMEs experienced the impact of the financial crisis differently from large firms. There are several reasons why SMEs could be more prone to experience difficulties in accessing bank credit, and more broadly, external finance, with the main reason being linked to specificities in their financing.[1] First of all, small firms are often believed to be more opaque and to have higher risk of failure than large firms. Second, small firms are more often young and have not had the time to build up a track record and reputation. Third, SMEs do not normally issue traded securities that are continuously priced in public markets so that to provide the market with information. From the bank's perspective (i.e. supply side), the costs involved in assessing and setting appropriate premia for risk and the relatively high monitoring costs may hinder the flow of funds to smaller firms.

In this chapter we investigate the issue of whether SMEs have suffered more than large firms using information directly provided by the firms through a new

[1] See for a review, ECB (2007).

A. Ferrando (✉)
European Central Bank, Frankfurt, Germany
e-mail: annalisa.ferrando@ecb.int

G. Calcagnini and I. Favaretto (eds.), *Small Businesses in the Aftermath of the Crisis*, Contributions to Economics, DOI 10.1007/978-3-7908-2852-8_9,

firm-level survey based on a sample of non-financial corporations in the EU (SAFE survey). General and firm-specific economic outlooks are largely blamed for this financing gap. But willingness of banks to provide loans is also highlighted.

We compare then the firms' view with the information gathered from the Eurosystem' Bank Lending Survey, which asks questions on business credit conditions to a sample of euro area banks. In this way we investigate the issue of whether banks had a more discriminating attitude in granting loans to smaller companies during the financial crisis. We complement the survey information with data on loan developments and interest rates charged by banks to non-financial corporations. While increases in risk premiums for SMEs are signalling a decline in credit supply, we cannot say whether this behaviour reflected justified credit risk considerations. However, the findings seem to point out that banks followed a more cautious lending policy towards younger and, hence, smaller firms, possibly because of their shorter track record and reputation. Since SMEs are often unable to switch from bank credit to other sources of finance, experiencing major financing obstacles can be a considerable challenge for them with respect to larger firms.

Finally, we report also the results of a more analytical exercise that investigates the underlying determinants of financing obstacles (measured through firms' self assessment of access to finance being the most pressing problem) among euro area firms and their implications for the use of external financing sources. The underlying data are related to the end of 2009 but the results hold also for subsequent waves of the SAFE survey. The main findings show that while the general sentiment of heightened financial obstacles was broadly based across firms during the recent crisis, age and ownership structure are important explanatory variables for firms' perceived financing obstacles across countries. As there is a strong relationship between age, simple structures of ownership and the size of firms, the results confirm the fact that SMEs were indeed hit harder when banks' credit standards tightened.

2 Financing Obstacle for SMEs During the Crisis

Our main source of information is derived from the ECB and European Commission survey on access to finance of small and medium-sized enterprises. The survey has been carried out four times between the summer of 2009 and March 2011. The survey contains firm-level information mainly related to major structural characteristics (size, sector, firm autonomy, turnover, firm age and ownership) as well as to firms' assessments of recent short-term developments regarding their financing needs and access to finance. A report containing the main results of the survey is published in the ECB website every 6 months.[2]

[2] For more information regarding the survey as well as the reports on the individual waves see http://www.ecb.europa.eu/stats/money/surveys/sme/html/index.en.html. See also ECB (2011).

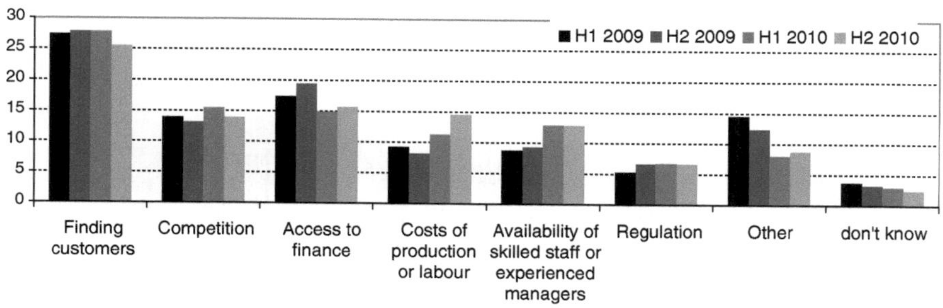

Fig. 1 Most pressing problems for firms in the euro area (Percentage of responding firms that used a specific type of financing over the previous 6 months) (Source: EU/ECB survey on access to finance. Notes: H1 broadly refers to the first half of the year while H2 to the second half. Please note that 2010H2 refers to September 2010 until February 2011)

Each surveyed firm is asked to identify the most pressing problem it is facing at the time of the SAFE survey.[3] It is therefore possible to identify a firm as being confronted with financing obstacles whenever it chooses "access to finance" as its most pressing problem. Since the beginning of 2009, the most pressing problem reported by euro area firms has been "finding customers", reported by nearly 30% of firms (Fig. 1). Clearly this factor reflects firms' concerns on current and future demand for their goods and services.

In the 2009 survey, "access to finance" came second in the implicit ranking of issues, with 19% of firms considering it to be the most pressing problem in the second half of the year. In the last two waves this share decreased slightly to 15%.[4]

As small firms tend to rely on credit obtained through intermediaries, the survey allows us to consider an alternative way to identify firms facing financing constraints based on their actual experience in applying for a loan. Indeed, respondents to the SAFE survey are being asked whether they have applied or not for a bank loan and whether they were successful in getting a loan. Based on this information, Fig. 2 shows that about 18% of SMEs experienced such constraints on bank loans in the second half of 2009 and 15% a year later. Interestingly enough, these figures are very similar to those based on the perceptions of financing obstacles.

[3] Each respondent is given a choice of seven alternatives: finding customers, competition, access to finance, costs of production (including labour costs), availability of skilled staff, regulation and other reasons.

[4] It is important to note that in SAFE we do not observe the *actual* levels of financing obstacles within a firm as well as whether access to finance is the second most pressing problem or the third most pressing, etc. (firms cannot signal more than one problem). In this way it could be that the survey underestimates the existence of firms that consider access to finance as a pressing (although not the most pressing) problem.

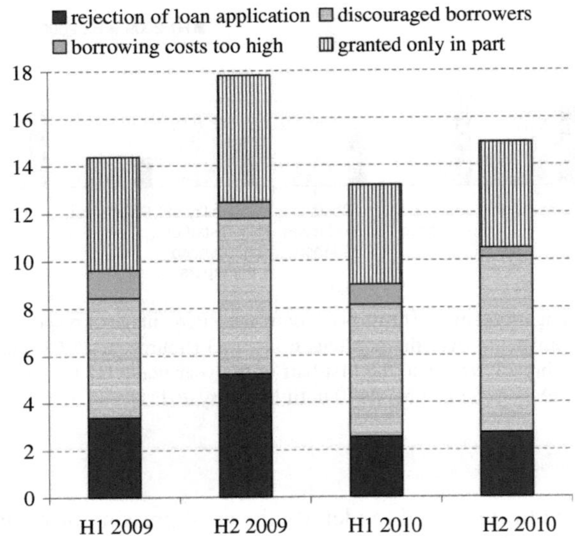

Fig. 2 Actual credit constraints of euro area SMEs (Source: EU-ECB survey on access to finance. Notes: The indicator is defined as the percentage of respondents SMEs who did not apply for a loan for fear of rejection; applied for a loan but saw their application rejected; had to refuse the loan because the cost was too high; or obtained less than the sum applied for. H1 broadly refers to the first half of the year while H2 to the second half. Please note that 2010H2 refers to September 2010 until February 2011)

3 Needs and Availability of Bank Loans for Euro Area SMEs During the Crisis

As seen in the previous section, a stable but relatively small proportion of firms were reporting access to finance as a major problem during the recent financial crisis but a larger one was pointing to difficulties to finding customers and to labour and production costs, hence to factors mainly related to the demand side. One question that could arise is then whether the strong decline in bank loans, which was witnessed in the euro area during 2009–2010, has really led to a reduction in financing means for small firms – hence the experienced difficulty of accessing finance – or whether this decline has been caused by reduction in demand for bank loans – as witnessed by the difficulties to finding customers causing a more cautions approach by firms about taking on debt.

As shown in Fig. 3, bank loans to corporations have indeed declined significantly during the financial crisis. In the period covered by the SAFE survey – between February 2009 and February 2011 – the rate of growth was on average growing 0.2% with the lowest level reached in the first quarter of 2010 at −2.5%. Since then, there has been a levelling-off of bank lending and a timid recovery. Among large

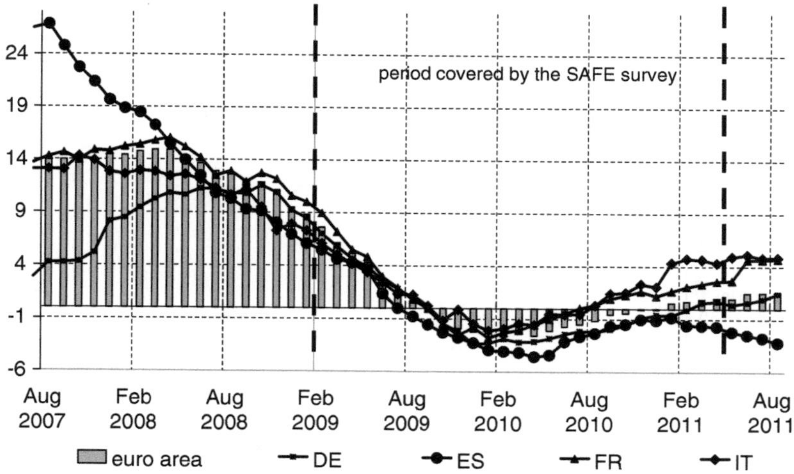

Fig. 3 Bank loans to non-financial corporations (annual rates of growth) (Source: ECB, MFI Balance sheet items. Note: the four waves of the SAFE survey cover the period between February 2010 and February 2011)

euro area countries, the decline in 2009–2010 has been particularly strong in Spain, as a result of the bursting of the private lending bubble in that country.

To understand whether the credit-related downturn led to a shrinkage in financing for SMEs, Fig. 4 summarises the interplay between needs and availability (an approximation of the *financing gap*) as reported by firms for the second half of 2009 (triangles in the chart), the first half of 2010 (circles) and the last part of 2010 until February 2011 (squares). In the lower right panel the financing gap becomes higher if availability of bank loans is deteriorating while needs for bank loans are increasing. This was generally the situation at the end of 2009 for most euro area countries, except for France. In the subsequent 6 months, bank loans needs were still increasing but at a slower pace, possibly driven by renewed efforts by firms to rely on internal financing. Moreover, there were some small signs of improvement in the availability of bank loans. As a result, the gap between needs and availability of bank loans was increasing at a slower pace with respect to the previous round (in graphical terms that corresponds to a general shift towards zero in the upper left panel). The last wave of the survey points to still existing but reduced credit flow bottlenecks with renewed increases in needs and still a deterioration in the availability of bank loans.[5]

[5] In this last round, French firms appear to have suffered a setback in accessing bank loans, especially compared with previous surveys. This development might be accounted for by the domestic debate taking place at the time of the interviews regarding the continuation of public guarantees and policy measures aimed at facilitating access to credit implemented during the crisis.

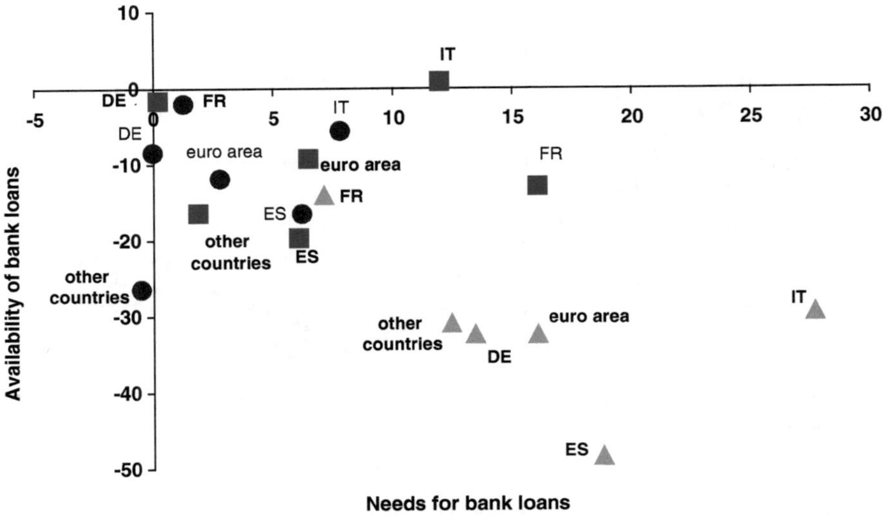

Fig. 4 Needs and availability of bank loans for SMEs according to the SAFE survey (net percentages) (Source: EU-ECB survey on access to finance and author's calculations. Notes: positive figures indicate that needs and availability are increasing with respect to the previous 6 months. *Triangles* indicate 2009H2; *circles* 2010H1 and *squares* 2010H2. In the lower right panel the financing gap is higher if availability of bank loans is deteriorating while the needs loans are increasing)

In most countries, SMEs perceived as the main factor affecting bank loan availability the general economic and firm-specific outlook (Fig. 5). Pure supply-side factors (i.e. captured by the willingness of banks to provide loans) are also considered to be a major factor. At country level, these factors have worsened further in Spain, while widespread improvements in the factors are reported in Germany.

Looking at the terms and conditions applied by bank when granting a loan to SMEs, firms reported an overall increase in price terms and conditions before the second half of 2009 and 2010 (Fig. 6). Non-price terms and conditions, which include collateral requirements, covenants and other guarantees, have generally been tightened. Developments across countries are more mixed. In Germany but also in Italy there was an improvement between the two surveys, while in France the percentages of SMEs indicating an increase in collateral and other requirements has risen compared with previous waves.

4 SMEs Financing During the Crisis: The Role of Banks

According to the credit-supply channel, a decline in credit in the banking system is often related to a more selective behaviour of banks in granting loans. Hence, given the characteristics of small firms of being more opaque and riskier entities than

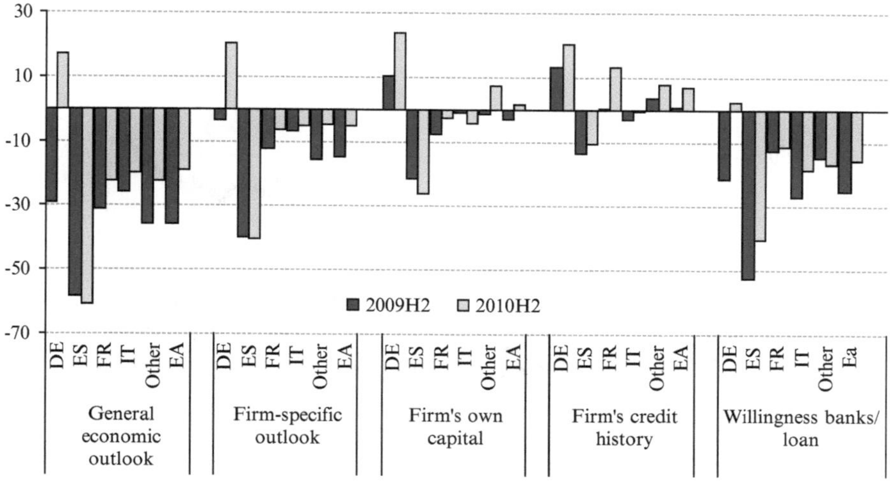

Fig. 5 Factors having an impact on the availability of external financing to euro area SMEs in 2009H2 and 2010H2 (net percentages) (Source: EU-ECB survey on access to finance. Note: a positive sign indicates that the factor has a positive impact on the availability of external financing)

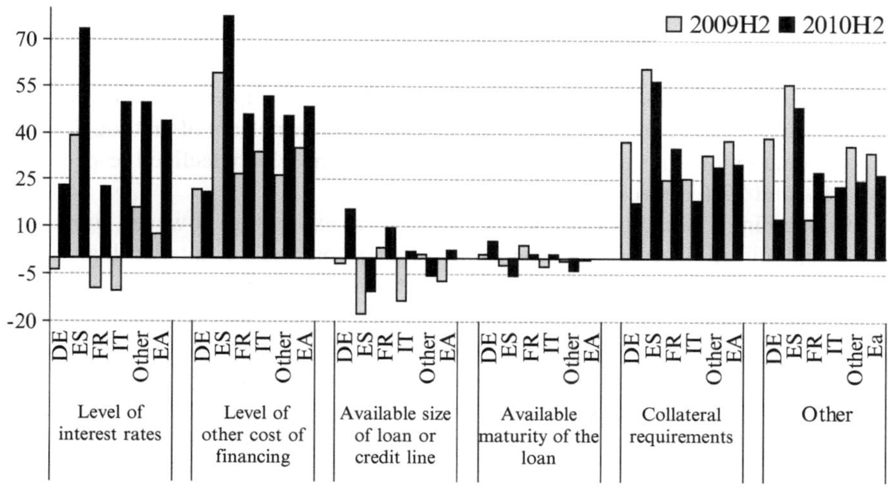

Fig. 6 Terms and conditions of bank financing for euro area SMEs in 2009H2 and 2010H2 (net percentages) (Source: EU-ECB survey on access to finance. Note: a positive sign indicates that the factor has increased)

larger firms, and given the fact that they are often unable to switch from bank credit to other sources of finance, it is more likely for SMEs to be affected by a credit crunch, or to experience lending obstacles as a considerable challenge for their day-to-day activities with respect to larger firms.

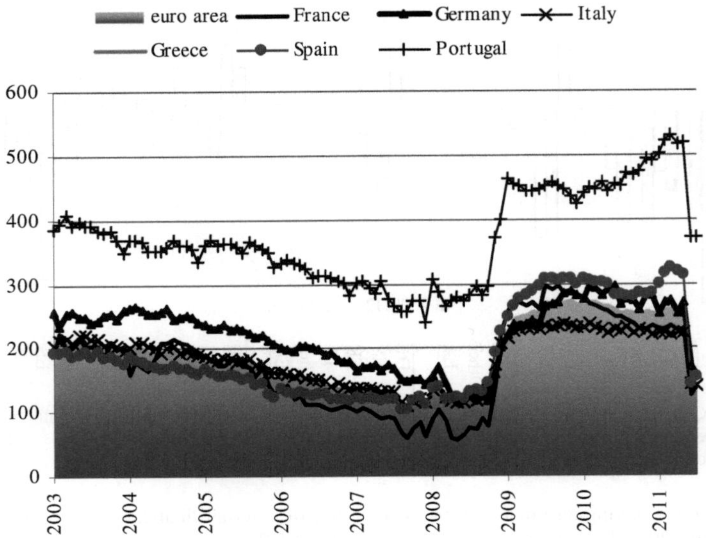

Fig. 7 Risk related premium on small-sized loans (in basis points) (Source: ECB, MIR Statistics. Note: risk related premium is defined as the difference between MFIs lending rates on short-term small-sized loans and 3-months Euribor)

Looking at development of bank flows, the general slowdown of lending activity in the euro area has concerned both small-sized loans (which are often thought as an approximation of loans to SMEs) and large-sized loans which declined respectively by 24% and 29% between 2008 and 2010.

Concerning lending pricing policy, banks were exhibiting, however, more prudent and discriminating risk managing in their pricing of smaller loans: The estimated risk premium on smaller loans (defined as the difference between short-term loans and the 3 month EURIBOR) in the euro area jumped from 133 basis points just before October 2008 to an average of 250 bps since May 2011 (Fig. 7). Also in countries under financial stress (Greece and Portugal), the risk-related premium on smaller loans has kept on increasing without a clear stabilization until May 2011, reaching 500 bps in that month. Since June 2011, the premium tended to decline mostly related to the increases in money market rates rather than to decline in lending rates.

The difference at euro area level between small-sized loans lending rates and large-sized loans also widened dramatically (from 76 bps before October 2008 to an average of 116 bps since then) (Fig. 8). The rise in the apparent discrimination towards smaller firms during the crisis is, however, not a common feature of all euro area countries. While the widening of the spread was very clear and large in France and Spain until mid-2009, and narrowed down since then, it was hardly detectable in Germany (or Portugal where, however the spread is on average the highest among euro area countries around 160 bps).

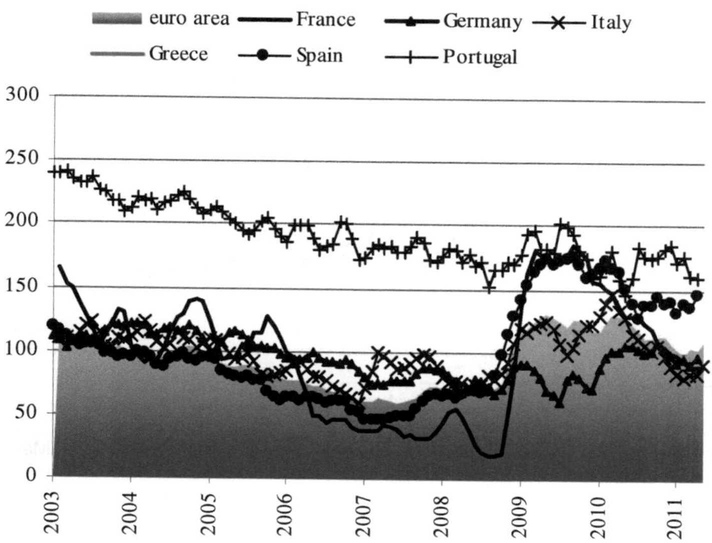

Fig. 8 Spread between small- and large-sized loan rates (3-months moving average; in basis points) (Source: ECB, MIR Statistics)

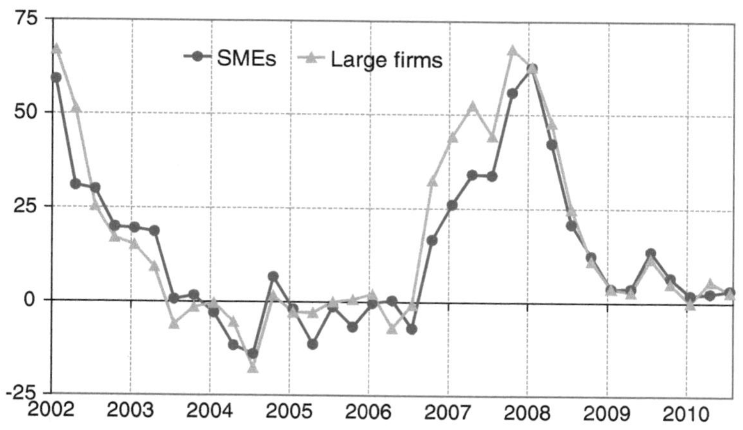

Fig. 9 Changes in credit standards on loans for SMEs and large firms (net percentage of banks contributing to the tightening standards) (Source: Eurosystem' s BLS and author's calculations)

An alternative way to investigate the issue of bank behaviour is to check what banks themselves report about their attitude towards SMEs using the results of the Eurosystem's Bank Lending Survey.[6] From Fig. 9, it is evident that banks tightened

[6] For more information on the bank lending survey and regularly published reports see http://www.ecb.europa.eu/stats/money/surveys/lend/html/index.en.html.

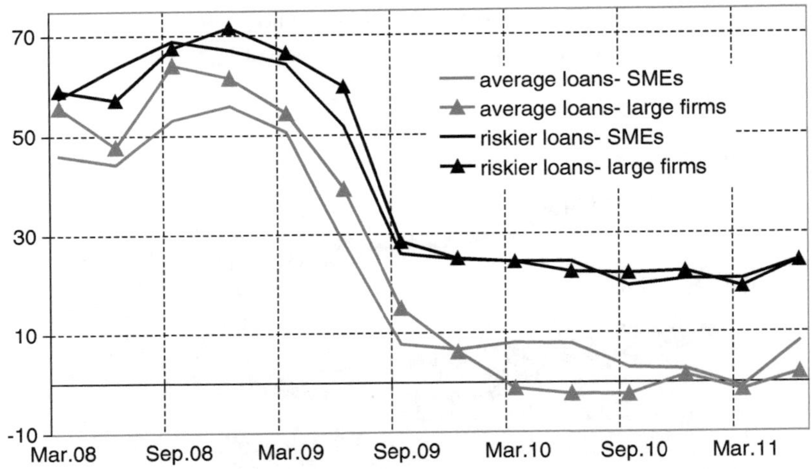

Fig. 10 Changes in margins applied to loans or credit lines for SMEs and large firms (net percentage of banks contributing to the tightening standards) (Source: Eurosystem's BLS and author' s calculations)

their credit standards at the onset of the financial crisis, reaching a peak at the end of 2008, but since returned to pre-crisis levels. An interesting piece of information for our investigation is that this has happened broadly independently of firm size, with larger firms being actually more penalized at the beginning of the crisis.

However, since 2010 banks reported a bigger widening of margins applied to those loans that are considered with no specific risk ("average loans") for SMEs than for large firms, while they seemed to have not discriminated according to size in the case of riskier loans (Fig. 10). As we do not have information from the survey about the quality of potential borrowers, we cannot say however whether these banking practices went beyond justified credit risk considerations.

Concerning riskier loans, it is expected that the bank's attitude towards riskier loans of SMEs is going to be changed in light of the new the new regulatory capital requirements set out in "Basel III".[7] According to banks' replies to the July 2011 bank lending survey, one fifth of the banks reported a decline in their risk-weighted assets for the past 6 months and expected a further decline for the next 6 months in order to comply with the capital requirements. This decline was – and is expected to remain – focused on riskier as opposed to average loans. Although in this first phase the adjustments were achieved mainly through tightening of credit conditions to large firms, banks foresaw loans to SMEs to be increasingly affected in the second half of 2011 and during 2012.

[7] See BIS (2010).

5 Determinants of Financing Obstacles for Euro Area NFCs

In this last section we report the results of an econometric analysis which investigates the underlying determinants of financing obstacles among euro area firms and their implications for the use of external financing sources.[8] We utilize the answers of almost 5,000 firms that participated to the second wave of the SAFE survey (H2 2009). The financing obstacles perceived by firms are defined as in Sect. 2 using the reply of firms to "access to finance" as their most pressing problem and, as shown in Fig. 1, this is the period when firms were reporting this problem with the highest percentage (19% in aggregate).

The probability of firms facing financing obstacles is modelled as a linear function of the characteristics available from the survey data (see Beck et al. 2006 and 2008; Coluzzi et al. 2009):

$$FinObst_{i,k} = \sum_k \theta_k country_k + \sum_j \phi_j (FirmCharacteristic_j)_{i,k} + \varepsilon_{i,k} \tag{1}$$

where 'FinObst' is the response by firm i in country k that indicates access to finance as most pressing problem. 'Firm Characteristics' is a vector of major firm attributes (e.g. firm size, sectoral dummies, firm age, and ownership structure). Country is a vector of country dummies to control for country-specific impacts on firms' responses. Given that the dependent variable is dichotomous, we use a probit model to estimate Eq. 1. We assume that the disturbance parameter, $\varepsilon_{i,k}$, has normal distribution and use standard maximum likelihood estimation. Since omitted country characteristics might cause error terms to be correlated for firms within countries, we allow for clustered error terms. Only observations with non-missing values on the relevant variables are considered, allowing the number of observations to be kept constant across estimations of the same type.

Figure 11 reports the percentage of firms in the sample that perceived access to finance being the most pressing problem for those characteristics that will be tested in the econometric analysis. Major heterogeneities are clearly related to the geographical environment and to the ownership characteristics of the firm. There are some indications of a negative relationship between age/size and perceived financing obstacles with older and large firms experiencing less financing obstacles.

Table 1 reports the results of a set of probit regressions with different combinations of those firm characteristics. The first model includes firm size, age, sector as well as country dummies. In the second model, the autonomy of firms is added while in the third model firm ownership is additionally taken into account. The last model finally considers interaction terms between size and ownership. In all specifications firm age seems to be one of the most robust predictors of firms' likelihood to experience financing obstacles. In particular, younger firms are more

[8] This section is based on the analysis carried out in Ferrando and Griesshaber (2010).

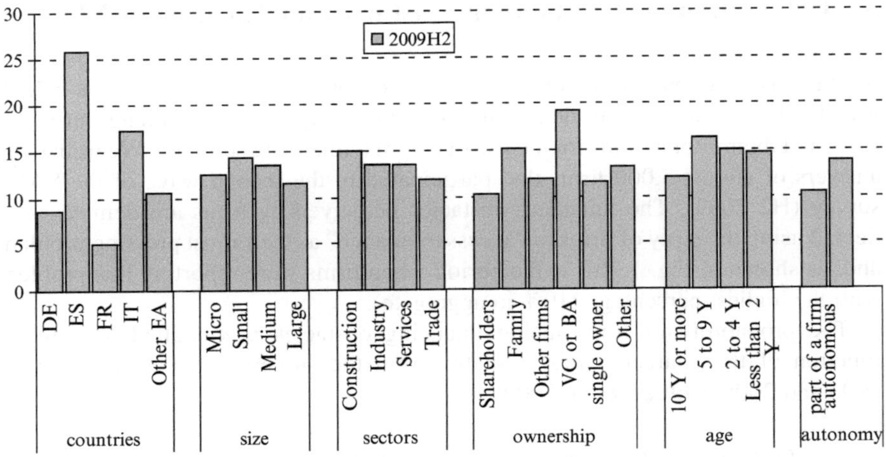

Fig. 11 Financing obstacles for euro area firms in 2009H2 (percentages) (Source: EU-ECB survey on access to finance. Note: VC stands fro venture capital and BA for business angels)

likely to experience problems in the access to finance. Turning to size, its effect does not prove to be robust across different specifications. Looking at the first specification, size seems to matter showing larger firms to have a lower probability of experiencing financing problems. Yet, the coefficient is significant only in the case of large firms. This effect is vanishing almost completely when financial autonomy of firms (column 2) and ownership structure (column 3) are included. However, although firm size has been identified as an important predictor of financing constraints by several empirical studies over the last years, these results are not necessarily contradictory to the existing research. For instance, using WBES data Beck et al. (2006) report that size matters for the overall sample of countries but becomes less important in developed countries. This is also confirmed for some euro area countries by Coluzzi et al. (2009) who show that size is not significant when a dummy on listed companies is included.

Turning to the results of the remaining covariates, financial autonomy does not seem to be of much importance. Although, firms that stand autonomous in their financial decisions are found to face a slightly higher probability of financing obstacles than firms which are part of groups either in the form of branches or subsidiaries (column 2), statistical significance remains very low and vanishes once ownership structure is controlled for (column 3). Rather more important, firms belonging to shareholders or to other firms suffer less than firms from the reference category, which groups together companies whose owners are a family, a group of entrepreneurs, venture capitalists, business angels or a natural person (one person only). However, this might be partially due to the fact that ownership structure is closely related to firm size as large firms are more likely to be listed on the stock market while smaller ones are in most cases owned by a family or a single person.

In all estimations we further control for sectoral and country specific effects. Contrary to most previous studies, sectoral differences are statistically significant.

Table 1 Experiencing financing obstacles – the role of firm characteristics

Dep var: Financing obstacles	(1)	(2)	(3)	(4)
Firm size[a]				
Small	−0.0188	−0.0116	0.00184	0.0248
	(0.122)	(0.123)	(0.124)	(0.134)
Medium	−0.0341	−0.00071	0.0425	0.117
	(0.0824)	(0.0889)	(0.0922)	(0.110)
Large	−0.149*	−0.0973	−0.0314	−0.0682
	(0.0784)	(0.0881)	(0.0892)	(0.134)
Log of firm age	−0.128***	−0.132***	−0.136***	−0.136***
	(0.0279)	(0.0293)	(0.0299)	(0.0305)
Sector[b]				
Construction	0.110	0.104	0.104	0.103
	(0.128)	(0.123)	(0.125)	(0.125)
Industry	0.0591	0.0665	0.0693	0.0695
	(0.0709)	(0.0759)	(0.0780)	(0.0794)
Trade	0.0523	0.0543	0.0475	0.0475
	(0.0374)	(0.0363)	(0.0370)	(0.0378)
Autonomous	–	0.216*	0.162	0.152
Profit oriented[c]		(0.110)	(0.114)	(0.111)
Firm Ownership[d]				
Shareholders, other firms or	–	–	−0.198***	−0.0177
Business associates, other			(0.0628)	(0.0567)
Small × Ownership				−0.173**
				(0.0827)
Medium × Ownership				−0.320***
				(0.102)
Large × Ownership				−0.0626
				(0.141)
Spain[e]	0.524***	0.537***	0.566***	0.563***
	(0.00892)	(0.0143)	(0.0149)	(0.0151)
France	−0.334***	−0.321***	−0.288***	−0.288***
	(0.00930)	(0.0146)	(0.0182)	(0.0183)
Italy	0.153***	0.159***	0.182***	0.185***
	(0.00588)	(0.00804)	(0.0128)	(0.0133)
Observations	4,742	4,742	4,742	4,742
Pseudo R^2	0.0645	0.0667	0.0697	0.0709

Notes: Probit regression results for access to finance being the most pressing problem as dependent variable (0, 1). Firm ownership is equal to 1 if the firm is listed or belongs to other firms or business associates and equal to 0 if it belongs to family, entrepreneur or single person, venture capitalist and business angel

Country dummies for remaining euro area countries are included but not reported. Only observations with non missing values on all relevant variables considered. Regression coefficients reported. Cluster robust standard errors in parentheses. Significance levels: $^{***}p < 0.1$, $^{**}p < 0.01$, $^{*}p < 0.05$

[a]Reference category is micro firms

[b]Reference category is service sector

[c]Reference category is part of a profit-oriented enterprise

[d]Reference category is family or entrepreneur, venture capital firm or business angel or natural person (one person only) as owner

[e]Reference category is Germany

Table 2 Firm characteristics and financing obstacles across countries

Dep var: Financing obstacles	Germany	Spain	France	Italy	Other euro area countries
Firm size[a]					
Small	−0.185	0.413***	−0.508***	0.135	−0.0607
	(0.129)	(0.121)	(0.177)	(0.124)	(0.120)
Medium	−0.0975	0.350***	−0.261	0.0608	0.0194
	(0.139)	(0.131)	(0.176)	(0.141)	(0.131)
Large	−0.0335	0.287	−0.239	−0.0127	−0.274
	(0.203)	(0.186)	(0.245)	(0.211)	(0.226)
Log of firm age	−0.135***	−0.243***	−0.123**	−0.0395	−0.0934*
	(0.0497)	(0.0564)	(0.0538)	(0.0499)	(0.0481)
Sector[b]					
Construction	−0.0299	0.361***	−0.394	−0.247	0.360**
	(0.160)	(0.132)	(0.252)	(0.193)	(0.153)
Industry	−0.0103	0.0552	0.289*	−0.182	0.385***
	(0.124)	(0.125)	(0.154)	(0.129)	(0.139)
Trade	−0.108	0.0411	0.0454	0.0330	0.381***
	(0.153)	(0.116)	(0.145)	(0.126)	(0.123)
Autonomous	0.180	0.395***	0.328*	−0.185	0.135
Profit oriented[c]	(0.192)	(0.138)	(0.186)	(0.155)	(0.141)
Firm Ownership[d]					
Shareholders, other firms,	−0.272*	−0.234**	0.131	−0.225*	−0.361***
business associates, other	(0.153)	(0.0992)	(0.132)	(0.120)	(0.124)
Observations	900	902	933	825	1,182
Pseudo R^2	0.0202	0.0485	0.0613	0.0142	0.0379

Probit regression results for access to finance being the most pressing problem as dependent variable. Only observations with non missing values on all relevant variables considered. Regression coefficients reported. Cluster robust standard errors in parentheses. Significance levels: $^{***}p<0.01$, $^{**}p<0.05$, $^{*}p<0.1$

[a]Reference category is micro firms

[b]Reference category is service sector

[c]Reference category is part of a profit-oriented enterprise

[d]Reference category is family or entrepreneur, venture capital firm or business angel or natural person (one person only) as owner

However, at this stage we are not able to infer whether this is attributable to the recent financial crisis – which could have been so widespread across sectors to possibly outweigh important sectoral differences with respect to the access to finance – or to the fact that our sample is indeed considerably different from those used in previous surveys. In contrast, country specific effects are found to be significant pointing to some cross-country heterogeneity in firms' financing obstacles. In Table 2 we present the estimated coefficients for Spain, France and Italy using Germany as reference category. The results indicate that Spanish, and to a lesser extent Italian firms, seem more affected by access to finance problems.

With respect to French firms, instead, the coefficient is negative indicating that they are significantly less likely to report financing obstacles than German firms.[9]

Finally, including interaction terms between size and ownership into the model, we reveal a negative and significant effect that (1) confirms the importance of the ownership structure and (2) partly explains the absence of significant effects for firm size in the previous specifications. Firms that are small or medium-sized in terms of employees and are owned by shareholders, other firms or business associates seem to possess a smaller likelihood of encountering financing obstacles than micro firms or firms of the same size classes but with different ownership. As expected, since most large firms in our sample are also public companies, there is no additional impact derived from the interaction terms when firms are large.

6 Conclusions

The most recent EU-ECB survey on access to finance confirmed that financial obstacles were one of the most cited factors impeding business dynamics. Comparing this piece of information with banks' lending pricing policy as well as their setting of credit standards applied to large and small firms during the crisis, we conclude that banks followed a more cautious lending policy towards younger and, hence, smaller firms, possibly because of their shorter track record and reputation. Since SMEs are often unable to switch from bank credit to other sources of finance, experiencing major financing obstacles has been a considerable challenge for them with respect to larger firms.

As a confirmation, we run an empirical investigation to determine the underlying determinants of financing obstacles and their implications for the use of external financing sources using the EU-ECB survey data for the second half of 2009.

The results indicate that while the general sentiment of heightened financial obstacles was broadly based across firms during the recent crisis, the firms that actually experienced financial constraints tended to be younger and, hence, smaller,

[9] Additionally, Ferrando and Griesshaber (2010) split the sample according to country and repeat the probit analysis including all major structural variables to investigate potential differences in the determinants of financing obstacles across countries. The results show that firm age again is a robust predictor except in the case of Italy. The same appears to hold for ownership with exception of France. Similar to the results from the aggregated sample, size is not statistically significant except in the case of Spanish small and medium firms which, according to the estimates, are suffering relatively more than micro firms from the access to finance as a pressing problem as well as in the case of small French firms for which the coefficient is negative. Tuning to the remaining covariates in the models, whereas almost no significant sectoral effects were found at the euro area level, things are slightly different when considering the major countries individually. In the case of Spanish firms, those active in construction sector are significantly more likely to experience financing obstacles than firms in the service sector. In France, a similar effect is found for manufacturing firms.

confirming the fact that SMEs were indeed hit harder through pricing and other terms and conditions when banks' credit standards also tightened.

Acknowledgments This chapter has benefited from discussions and comments from I. Ganoulis, D. Rodriguez Palenzuela as well as from participants at the Conference "The Economics of Small Businesses in the aftermath of the crisis. Cross-country Analyses and Policies" (University of Urbino). The views expressed are those of the author and should not be attributed to the European Central Bank.

References

Bank for International Settlements (2010)Basel III: a global regulatory framework for more resilient banks and banking systems. Basel Committee on Banking Supervision, 16 Dec 2010 (http://www.bis.org/publ/bcbs189.pdf)

Beck T, Demirgüç-Kunt A, Laeven L, Maksimovic V (2006) The determinants of financing obstacles. J Int Money Finance 25:932–952

Beck T, Demirgüç-Kunt L, Maksimovic V (2008) Financing patterns around the world: are small firms different? J Financ Econ 89:467–487

Coluzzi C, Ferrando A, Martinez-Carrascal C (2009) Financing obstacles and growth: an analysis for euro area non-financial corporations. ECB WP No. 997

European Central Bank (2007) Corporate finance in the euro area, ECB occasional paper series, No. 63

European Central Bank (2011) Financing obstacles faced by euro area small and medium-sized enterprises during the financial crisis. Financial stability review, June

Ferrando A, Griesshaber N (2010) Financing obstacles among euro area firms: who suffers the most? ECB working paper series, No. 1293

Examining the Impact of Credit Access on Small Firm Survivability

Traci L. Mach and John D. Wolken

1 Introduction

The recent economic turmoil beginning in late 2007 has challenged businesses of all sizes. Firms have been faced with a great deal of uncertainty regarding sales and the economic outlook. At the same time, the recent downturn has dramatically impacted the availability and terms of credit. Over the 2007–2009 period, financial institutions have reported tightening their credit standards for approving loans (SLOOS).[1] Many small business owners rely on personal assets to guarantee or collateralize loans for their firms. As their equity in real estate holdings has generally declined in value during the recent turmoil, owners' ability to tap into personal balance sheets to secure their business credit needs has also shrunk (NFIB 2010).[2]

This paper examines the effects of credit availability on small firm survivability over the period 2004–2008 for non-publicly traded small enterprises. We develop failure prediction models for a sample of small firms that were confirmed to have been in business as of December 2003, with particular attention to the impact of

The disclaimer applies to both authors. The views expressed herein are those of the authors. They do not necessarily reflect the opinions of the Federal Reserve Board or its staff. The authors would like to thank participants of the Financial Structure, Financial Studies, and Microeconomics Surveys Sections' brown bag workshop for helpful comments and suggestions on an earlier version of this research. Valuable research assistance was provided by Timothy Satterfield and Samantha Dalton.
John Wolken was formerly a member of the staff of the Board of Governors.

[1] Senior Loan Officer Opinion Survey on Bank Lending Practices http://www.federalreserve.gov/boarddocs/SnloanSurvey/

[2] National Federation of Independent Businesses *Small Business Economic Trends* http://www.nfib.com/tabid/83/Default.asp. See also Avery et al. (1998).

T.L. Mach (✉) • J.D. Wolken
Board of Governors of the Federal Reserve System, Washington, DC, USA
e-mail: traci.l.mach@frb.gov; john.d.wolken@gmail.com

G. Calcagnini and I. Favaretto (eds.), *Small Businesses in the Aftermath of the Crisis*, Contributions to Economics, DOI 10.1007/978-3-7908-2852-8_10,

credit constraints.[3] We find that credit constrained firms were significantly more likely to go out of business than non constrained firms. Moreover, credit constraint and credit access variables appear to be among the most important factors predicting which small U.S. firms went out of business during 2004–2008 even though an extensive set of firm, owner, and market characteristics were also included as explanatory factors.

Baseline data are taken from the 2003 Survey of Small Business Finances (SSBF) and provide information on the balance sheets, credit use, and credit constraints faced by a representative sample of more than 4,000 small (fewer than 500 employees) firms that were in operation at the end of 2003. The SSBF allows construction of standard financial balance sheet and income ratios that are used in many failure prediction models (Altman (1968)).[4] In addition to financial ratios, the SSBF data are rich in firm (e.g., employment, organizational form, location, age and industry) as well as owner characteristics (e.g., race, education, experience, home ownership and value of home). Importantly for this study, the data also contain many measures of credit constraint, including a firm credit score, self-reported credit history variables such as late payments and bankruptcy for firm and principal owner, recent credit application experience of the firm, and whether firms borrow using trade credit and credit cards.

Our measure of survivability is constructed from the NETS database from Walls & Associates (2009).[5] The NETS database takes cross-sectional data from the Dun & Bradstreet Market Identifier File and constructs a panel of sales and employment over time. These data allow us to track which firms closed or went out of business, were absorbed by another firm, or continued operations in each year subsequent to 2004. We estimate a reduced-form logistic model of a firm being in business in 2008. The results indicate that credit access and credit quality are significant indicators of the firm still being in business. We then estimate a proportional hazard model with yearly indicators of the firms' operational status. The hazard model further underscores the importance of credit access and quality in the ability of the firms to continue operations.

2 Previous Literature

The literature on default prediction studies is voluminous and involves several approaches.[6] The earliest, most well-known and widely applied technique to firm failure prediction models uses financial ratio analysis. Following the introduction of

[3] More precisely, to be eligible for the survey, a firm had to have been in business at the end of 2003 and at the time of the interview which was conducted during 2004. For more information, see Mach and Wolken 2006.

[4] Typically, balance sheet or income information is not available for small firms.

[5] http://www.youreconomy.org/nets/?region=Walls

[6] For summary articles, see Hall (1992) and Keasey and Watson (1991).

such models in the 1960s, researchers provided theoretical arguments for failure that tended to argue one or both of the following hypotheses. The first is that firms will fail because the present value of their costs exceeds their revenues. The second is that inefficiencies in capital markets can lead to failure among firms with positive net present values. While financial statement data are required for ratio analysis studies, the theoretical literature on firm failure has suggested other variables, including firm and owner characteristics other than financial statement variables, local and macro economic conditions, as well as indicators of credit access and credit quality. The various approaches will be discussed below.[7]

2.1 *Financial Ratio Analysis*

Studies predicting failure of businesses began with work in the 1930s that established that accounting ratio measures exhibited by discontinuing firms were different from measures exhibited by surviving firms (Shailer (1989). Following up on this finding, Beaver (1966) and Altman (1968) explored the differences on small samples of failed and non-failed businesses. Beaver examined some 14 financial ratios individually on a matched sample of 79 failed and 79 non-failed institutions. Altman (1968) is often credited as the first to apply multivariate techniques to failure analysis. He applied multiple discriminant analysis to a sample of 33 failed and 33 non-failed institutions that filed bankruptcy petitions during the period 1946–1965. From an initial 22 ratios, Altman settled on five ratios as providing in combination the best overall prediction of corporate bankruptcy. The ratios included measures of liquidity, profitability, leverage, solvency and activity ratios.

Many studies since Altman (1968) have added further support to the potential usefulness of such models for failure prediction. The literature is quite large and studies that provide reviews of earlier works include Ohlson (1980), Taffler (1982), and Altman and Sabato (2007). Such studies have used various statistical techniques (e.g., multiple discriminate analysis, logistic and probit analysis, and factor analysis) and have been applied to several countries including the United States and Great Britain. It is fair to say that most such studies have been concerned with large publicly traded (listed or quoted) firms – in part because financial statements for such companies are generally available.

Because the focus has been on larger concerns, samples of failed enterprises in these studies tend to be quite small. The early work generally drew matching samples of non-failed companies, whereas more recent studies have included far larger samples of non-failed institutions. For example, Altman and Sabato (2007)

---

[7] In addition to studies predicting failure or discontinuance, there are also studies of entry and exit rates among businesses. The entry-exit studies provide some useful insights into reasons underlying failure, even though such studies may be largely descriptive. See Hall (1992), p. 239, and also Dunne et al. (1988), Headd (2003), Hudson (1986), and Boden (2000).

employ a sample of about 2,000 firms between 1994 and 2002. However, only 120 observations were observed failures.

Little attention was initially given to small or private enterprises. Publicly available datasets are heavily biased towards large firms and usually contain little or no information on small firms. However, some efforts to study small firm failure have produced encouraging results, although often the variables and models used for small enterprise studies differ from the ratio analysis described above. This is in large part because financial statement data for smaller enterprises or for non-publicly traded enterprises is often unavailable.[8] Studies of private companies using financial ratio analysis are reviewed by Shailer (1989). With the exception of Edminster (1972) – who used a sample of 21 firms reporting losses and 21 non-failed businesses drawn from the US Small Business Administration borrowers for 1954–1969 – these studies have generally analyzed private United Kingdom (Briggs and MacLennan (1983), Peel and Peel (1987)) or Australian companies (Shailer (1986)). Samples were typically on the order of 100 or less observations. In- sample predictions averaged around 75% accurate. But when applied to a hold-out sample, Peel and Peel's models exhibited high classification error rates. Differences in the variables used between small (private) company models and large (public) company models as reported in these studies suggest that small business failure models may be a separate avenue of research.

2.2 *Net Present Value*

Employing the hypothesis that survival is dependent upon a positive net present value (of revenues and costs), several studies have identified firm characteristics other than financial ratios that are likely associated with increasing the difficulty firms have in maintaining solvency. Models formulated by Argenti (1976), Jovanovic (1982) and Dunne et al. (1988) indicate that firms fail because of their management structure or because they are unlucky or inefficient. Argenti (1976) viewed financial ratios as symptoms of failure, not the cause.[9] Rather, the process of failure is based on a number of defects in the organization and financial structure of

[8] There are ratio studies that claim to have examined small firms. However, most of these studies use publicly available information which generally includes firms that are publicly traded (or "quoted" or "listed" or "not private"). For example, Altman and Sabato (2007) wrote on failure prediction for SMEs. In their study, an SME is a firm that has less than $65 million in sales but is available on Compustat. Compustat in 2007 contained about 24,000 listings. There are approximately 24 million enterprises in the United States.

[9] Credit quality or credit access may also be a symptom, rather than a cause if it is firm actions (risk taking, performance, etc.) that result in the credit constraints. On the other hand, credit constraints resulting from frozen credit markets or deterioration of the financial institution used by the firm may be causal. Future research on this topic will attempt to sort out the nature of the credit constraints and their effect on survivability.

the company. Based on case studies, Argenti concludes that factors reflecting the management structure and the adequacy of the accounting information system such as one-man rule, lack of management depth and experience, and concentration of resources into a single project are factors that are associated with failure. Although not specifically examining small firms, these factors are usually characteristics of small firms. Studies that included some of these factors include Storey et al. (1987) and Keasey and Watson (1986). Both studies included financial ratios as well as some variables reflecting management structure. They found that the nonfinancial ratio variables increased the predictive accuracy somewhat. The results are generally supportive of the Argenti model of business failure. Such variables also have some practical advantages in that some of the information required is available for small firms, whereas financial statement variables are difficult to obtain for non-traded enterprises.

Jovanovic provides a two period model of firms' costs and Dunne et al. test some of the implications to further explore why failures are more likely among small and young enterprises. In these models, small and young firms are likely to be more vulnerable because it takes time to acquire the information and knowledge to better predict their costs and revenue functions than it does for larger and older firms.

Hall (1992, p. 240) further argues that small firms have limited portfolios – first with respect to the opportunity-sets of products or services they offer or markets in which they operate, and second with respect to the human capital embodied in management, which is likely less vis-à-vis larger firms. Both the degree of diversification and level of human capital will likely increase with the size of the company. Consequently, small firms are more prone to failure. Limited portfolios will lower the expected earnings and increase the variance of small vis-à-vis larger firms. And a result of a lower volume of human capital is that the income streams that it generates may be both lower and more sensitive to the situation being faced. These observations reinforce the importance of accounting for shocks to the firms' operation. But they also suggest that the characteristics of management – including factors such as experience, education, and breadth of management – may be important to identifying firms likely to fail.

2.3 Capital Market Imperfections, Credit Constraints, and Credit Quality

Another line of research focuses on capital market imperfections (Wadwhani (1986), Hudson (1986), Simmons (1989)), and by implication credit constraints and credit quality. Inefficiencies in capital markets can lead to failure among firms with positive net present values. In particular banks (and other lenders) have been accused of charging small firms too much, demanding too high a level of collateralization, being inefficient in their procedures for credit assessment (or economies of scale with firm size), and in some cases unwilling to lend to smaller firms.

As a result of these imperfections, small firms facing such constraints will be less able to adjust to shocks to their operations than larger firms.

Early tests of failure models based on capital market imperfections use variables that were intended to be surrogates for shocks to cost or revenue flows, as was true of the studies by Dunne et al. (1988), Storey et al. (1987) and Keasey and Watson (1986). Hudson includes a variable of company profits to company GDP which might be considered a surrogate for net present values; Wadhwani does not include inflation but does find that interest rates are positively correlated with the numbers of liquidations.

The literature also suggests several indicators of credit access and credit quality that are likely related to the probability of discontinuance. Applying and being denied credit signals that a firm is unable to secure the desired level of credit. Credit constrained firms have less access to credit which may also force them to use relatively more expensive shorter term debt to finance their operations. Such restrictions and costs may leave them vulnerable to shocks in their cash flow. This in turn puts them at a higher risk of failure (Keasey and Watson 1991). It may also constrain the firm from initiating new projects, expanding operations, investing in research and development, or even making its payroll, and ultimately affect the firm's probability of survival. For example, Musso and Schiavo (2008), report a significant relationship between survival and financial constraint using an index of financial constraint.

Such constraints may cause firms to turn to alternative but somewhat more expensive types of credit. For example, borrowing using credit cards or using trade credit are often more expensive than traditional bank loans. Blanchflower and Evans (2004) found that firms that had their credit access constrained–i.e., denied credit or feared applying because they thought they would be denied–were significantly more likely to borrow using credit cards. In a recent study of start-up firms, Scott (2009) reports a negative correlation between the level of credit card debt and subsequent survival. And Peterson and Rajan (1997) report that firms that had loans requests denied were more likely to borrow using trade credit.

In addition to credit constraint indicators, the credit quality of firms may also be related to subsequent survival. Several studies have found that firms with poor credit histories (e.g., bankruptcy, delinquency on current debt, and judgments), or low credit scores, are less likely to have loan requests approved (e.g., Blanchflower and Evans. (2004), Cavalluzzo and Wolken (2005), Hyytinen and Pajarinen (2007). Robb and Robinson (2010) examine differences between surviving and failing "new" firms and report that firm credit scores are positively associated with indicators of success – revenues, assets, profits, and employees larger than the sample median levels.

In sum, the literature regarding predicting business failure has used a variety of approaches, starting with financial ratio analysis in the 1960s. Since then, researchers have identified a number of other factors likely to affect the probability of failure, in part due to the development of theoretical models to help understand what leads to failure and in part due to necessity owing the lack of publicly available financial data for small firms. The literature indicates that factors that are associated with (small) firm insolvency include financial ratios as well as other income and balance sheet variables such as size. Moreover, other firm characteristics (e.g., age) and owner/manager characteristics (experience, education) are likely important, as are variables reflecting credit constraints and credit quality of the firm. Finally, given that firm performance is likely to be affected to some degree by the general health (or change in the health) of the local and national economy, it may be important to include such variables as well.[10]

3 The Data

3.1 2003 Survey of Small Business Finances (SSBF)

The majority of data for this paper come from the 2003 Survey of Small Business Finances (SSBF). The 2003 SSBF was conducted to collect information from the owners of a nationally representative sample of more than 4,000 U.S. small business enterprises. Owners were asked about firm income statements; balance sheets; financial relationships; credit experiences; lending terms and conditions; the number of branches and firm headquarters location; the types of, and locations of financial institutions that were used; and about various other firm characteristics.[11]

The target population of the survey was defined as for-profit, non-governmental, non-depository and non-agricultural enterprises with fewer than 500 employees. Firms in the sample had to be either single establishments, or the headquarters of multiple establishment enterprises that were not majority owned subsidiaries of other firms. Additionally, in order to be eligible, firms had to have been in business during December of 2003 and at the time of the interview. The majority of interviews occurred between June and December of 2004.

The Dun and Bradstreet (D&B) Market Identifier file (DMI) was used to construct the sampling frame for the 2003 SSBF. The DMI contains minimal basic company data on U.S. businesses. It is a meant to be a snapshot of active businesses *at a particular point in time*.

3.2 National Establishment Time-Series (NETS) Database

In order to incorporate a time element into the SSBF cross-section, we merge in data from the National Establishment Time-Series (NETS) Database. The NETS Database is constructed from 20 "snapshots" taken every January since 1990 of all

---

[10] See Holman and Fletcher (1989). They examine factors affecting failure rates and find that macroeconomic factors such as the money supply, real GNP, and real corporate profits are highly significant in predicting firm failure.

[11] For detailed information about the 2003 SSBF, see 2003 Methodology Report (National Opinion Research 2005). Selected survey results are summarized in Mach and Wolken (2006). The database includes a credit score for each firm purchased from Dun and Bradstreet.

active Dun and Bradstreet establishments (currently over 24 million). Each year, the snapshot is compared to information from the previous years. No establishments are ever deleted from the database; but their last year in business is indicated making it possible to follow a firm over time. Walls & Associates maintains the NETS database and continues to update estimates before the next annual release.[12]

One unique feature of the NETS database is that it allows the researcher to distinguish between a firm that is no longer in business and one that has been purchased by another firm or changed its organizational form but essentially continues to operate. On the D&B DMI file, firms that were purchased by other firms or changed organizational form are assigned a new DUNS number. Thus, comparing two point-in-time snapshots of the D&B file without the quality control implemented for the NETS database would make these firms that went out of business appear identical to firms that were still operating with different names or forms. However, the NETS database does not allow us to differentiate between firms that went out of business due to bankruptcy or poor performance from those that closed down because the owner simply retired.[13]

The NETS database also contains several other yearly measures of firm characteristics such as employment, sales and credit ratings. Because our experience with the employment and sales numbers from the DMI file have indicated that these numbers are not very accurate, we do not use this information in analysis.[14] However, because D&B specializes in credit rating, this measure may be more informative. The credit rating is a composite credit worthiness measure; unlike the credit score (the measure available on the SSBF dataset), which provides a *relative* ranking of firms, the credit rating (the measure available on the NETS database) is *specific* to the firm.[15]

[12] For more information on the construction of the NETS database, see Walls & Associates (2009). As new data becomes available (the next year), Walls and Associates reexamines and corrects, if necessary, previous years' data as well. The results reported in this paper are for the NETs database as of 2009. Due to the dynamic nature of the database, results may be slightly different if the models were to be estimated with a different release of the NETs database even if the same time period were examined.

[13] The models control for this possibility by including a variable on the owner's age. See below.

[14] For example, the correlation coefficient of total sales provided on the D&B DMI and total sales provided by the individual firms is 0.6 and the correlation coefficient of employment is similar.

[15] A detailed interpretation of the credit rating can be found at http://www.dnb.com/about-dnb/15062603-1.html. The variable is a multi-tiered variable based both on the firm's credit history and size. For our analysis, we construct a discrete variable with five categories: high, good, fair, limited, or not rated. Its value may change from year to year. This is separate from the credit score measure that ranks all firms from 1 to 100, with 1 being the most risky firm. The credit score is only available for 2003.

3.3 *House Price Index and Geographic Controls*

Many small business owners rely on personal assets to guarantee or collateralize loans for their firms. As their equity in real estate holdings has generally declined in value during the recent turmoil, owners' ability to tap into personal balance sheets to secure their business credit needs has also shrunk (National Federation of Independent Businesses 2010). In order to account for this shrinking source of credit, we include the change in Federal Housing Finance Agency (FHFA) house price index for the area where the firm's headquarters are located in the model.[16]

Because the location where the firm operates is likely to have an impact on the survivability of the firms, we also include a limited number of measures of the economic environment. Using the MSA or rural county where the firm's headquarters is located, we merge in measures of population and establishment density from Census data, the average wage per job and per capita income from BEA, and the unemployment rate from BLS. As an area becomes more densely populated, we would expect the firm to have more demand for its products, increasingly the likelihood that the firm will stay in business. The relationship with the number of establishments in the area is theoretically less clear. While higher densities of businesses would likely be associated with better infrastructure decreasing the cost of doing business, more businesses could also mean more competition. Because small businesses are also potential employers, rising wages in the area where the firm operates is likely to increase the cost of production; thus, we would expect to see an inverse relationship between average wages and firm survivability. Per capita income, on the other hand, should have a positive relationship with survivability as more wealthy residents would be more capable of buying the firm's output. We would expect unemployment rates to be negatively associated with survivability.

To capture the changing nature of the economic environment and home prices, we use the percentage change in these measures between 2004 and 2008 when we estimate the logistic models and the 1-year lagged value (time-varying) when we estimate the proportional hazard model.

4 The Sample

Of the 4,240 firms that completed the 2003 SSBF, 4,230 were successfully matched to firms on the NETS database.[17] Of those firms, 332, just under 8%, were no longer in business as of January 2008 (Table 1). This represents slightly less than 10% of

[16] For more information on the construction of the FHFA index, see Calhoun (1996). We also employed an alternative index – the LoanPerformance index – with similar results. For additional details, see Mach and Wolken (2011).

[17] The ten firms that did not match up are still being investigated. The most likely explanation is that the D&B number was incorrect on either the SSBF sample control file or the NETS database.

Table 1 Firms out of business, by last year in business

Last year in business	Unweighted frequency	Unweighted percent	Weighted percent
2004	45	1.06	1.25
2005	104	2.46	2.90
2006	109	2.58	3.21
2007	74	1.75	2.18
Total	332	7.85	9.54

Table 2 Sample restrictions by implicate

Implicate	Dropped sample			Sample used in estimation		
	Assets < \$0 or sales < \$1,000	Still in business in 2008	Not in business by 2008	Total	Still in business in 2008	Not in business by 2008
1	162	138	24	4,068	3,760	308
2	161	137	24	4,069	3,761	308
3	162	137	25	4,068	3,761	307
4	158	135	23	4,072	3,763	309
5	160	135	25	4,070	3,763	307

the population. Statistics on new employer firms indicate that 69% survive at least the first 2 years and 51% survive at least 5 years (Small Business Administration 2009). Because the 2004 SSBF firms are not all new firms – the median firm was 12 years old – and they are not all employer firms, one would not expect the survival rates to match perfectly, but the 8% does not seem extraordinarily high or low.

In order not to bias the results, we did not want to include firms in the analysis that already looked defunct when they were interviewed in 2004. For the analysis, we eliminated firms that reported assets of less than \$0 or sales of less than \$1,000. Because the SSBF were multiply imputed, this restriction led to slightly different numbers of firms being used for each implicate. Table 2 provides the breakdown of the sample by implicate. Between 158 and 162 firms were dropped from analysis in each implicate. About double the fraction of these marginal firms were no longer in business in 2008 than the other firms. The analysis that follows was conducted using the remaining firms in the 2003 SSBF.

5 Descriptive Statistics

Table 3 provides some descriptive statistics of the sample by whether or not the firm was still in business by 2008.[18] A quick overview of the data shows that the firms that were still in business by 2008 looked different than those that were not when

[18] Descriptive statistics and model estimates are based on the fully imputed data with standard errors adjusted to take into account the multiple imputations; they are also weighted to take into account sample design. Definitions of variables are summarized in Mach and Wolken (2011).

Table 3 Means and medians by out of business status in 2008

	Out of business by 2008		Still in business in 2008	
	Mean	Median	Mean	Median
Credit quality variables				
Firm credit score	45.29***	42.00	53.20	58.00
Firm or owner declared bankruptcy	0.04	0.00	0.03	0.00
Firm or owner delinquent 3+ times	0.14**	0.00	0.09	0.00
Judgment against firm or owner	0.09**	0.00	0.04	0.00
Had no D&B credit rating in 2003	0.94***	1.00	0.78	1.00
Credit access variables				
Firms fearing denial	0.23	0.00	0.17	0.00
Borrowing on credit cards	0.30**	0.00	0.22	0.00
Borrowing on trade credit	0.33**	0.00	0.24	0.00
Been denied trade credit	0.06	0.00	0.05	0.00
Used real estate as collateral	0.24	0.00	0.20	0.00
FHFA home prices (% Δ 2004–08)	18.72	18.10	19.08	18.60
Firm/owner variables				
Firm age	12.55***	10.00	14.65	12.00
Firm employment	7.94	3.00	9.00	4.00
Log of firm assets	11.27	11.28	11.21	11.23
Log firm sales	12.18	12.20	12.26	12.26
Firm has zero or negative equity	0.19	0.00	0.19	0.00
Older than 60 indicator	0.19	0.00	0.20	0.00
More > median experience	0.37	0.00	0.42	0.00
Financial ratios				
Liabilities/assets ratio	0.96*	0.33	1.83	0.28
Cash/sales ratio	0.20*	0.04	0.13	0.04
Accounts payable/sales ratio	0.04	0.00	0.05	0.00
Pretax profit/assets ratio	6.93	0.22	4.66	0.34
Pretax profit/(liabilities†) ratio	25,411.04	1.08	15,177.11	1.04
Division variables				
New England	0.11**	0.00	0.15	0.00
Mid Atlantic	0.06	0.00	0.05	0.00
East N Central	0.09***	0.00	0.14	0.00
West N Central	0.13**	0.00	0.07	0.00
South Atlantic	0.04*	0.00	0.06	0.00
East S Central	0.20	0.00	0.16	0.00
West S Central	0.19	0.00	0.19	0.00
Mountain	0.07	0.00	0.07	0.00
Pacific	0.13	0.00	0.10	0.00
Industry variables				
Construction and mining	0.15	0.00	0.12	0.00
Manufacturing	0.07	0.00	0.07	0.00
Transportation	0.03	0.00	0.04	0.00
Wholesale trade	0.05	0.00	0.06	0.00
Retail trade	0.21	0.00	0.18	0.00
Insurance agents and real estate	0.09	0.00	0.07	0.00
Business services	0.21	0.00	0.25	0.00

(continued)

Table 3 (continued)

	Out of business by 2008		Still in business in 2008	
	Mean	Median	Mean	Median
Professional services	0.19*	0.00	0.21	0.00
Business or professional services	0.39*	0.00	0.46	0.00
Market condition variables				
Population density (% Δ 2004–08)	178.85*	145.93	149.76	98.56
Average wage (% Δ 2004–08)	16.53	15.89	16.10	15.69
Unemployment rate (% Δ2004-08)	6.31	2.96	5.88	5.25
Establishment density (% Δ 2004–08)	4.94	5.33	4.30	3.85
Per capita income (% Δ 2004–08)	20.10	20.11	20.24	20.03

Notes: Means and medians are based on the firms that were matched to the NETS database and had assets greater than zero and sales of more than $1,000 when interviewed in the 2003 SSBF and have been weighted to account for sampling. Standard errors for a test of differences of means between the out-of-business and still-in-business populations have been adjusted to account for the multiple imputation

The difference of means is * significant at 10%; ** significant at 5%; *** significant at 1%. † indicates $1 was added to the variable to avoid $0 in the denominator of logged value

they were interviewed in 2004. In terms of credit history, the firms that were no longer in business in 2008 had credit scores that were 8% points lower than their counterparts. In 2003, the majority of small firms were not rated by D&B – an indicator of the opacity of small firms. However, firms that were no longer in business by 2008 were much more apt to have not been credit rated by D&B, likely making it more difficult for potential lenders to evaluate their credit worthiness. The discontinued firms were also more likely to report that the owner or the firm had been 60 or more days delinquent in paying three or more bills in the past 3 years or had a judgment rendered against them.

Discontinued firms were generally smaller than continuing firms, and reported smaller amounts for both balance sheet and income items: Discontinued firms had smaller sales and profits; they had lower liabilities and accounts receivable as well. As a baseline comparison to the early failure prediction literature, we construct the financial ratios of leverage, liquidity, activity, profitability, and coverage as suggested by Altman.[19] There were large differences in the average ratios across the two groups, although few differences are statistically significant. For instance, discontinued firms reported average pretax profit to liabilities ratios above 25,000 while continued firms reported an average near 15,000. However, their medians were quite similar. This gives some idea of how heterogeneous the finances of small businesses are and possibly some insights into the potential (or lack thereof) of financial ratios to predict the discontinuance of very small businesses.

[19] We constructed multiple measures of each of the components, but only include one of each such measures in the analysis. Summary statistics for the other measures are provided in Mach and Wolken (2011).

In terms of firm demographics, the discontinued firms tended to be younger and smaller, although the differences in size measures (assets, sales, profits, and employment) other than profits are not statistically different. There is virtually no difference in forms of organization. There are only small differences in the characteristics of the owners of the firms that were discontinued; they were slightly more likely to be minority owned and had less formal education and less business experience.

6 Econometric Results

The data contain five completely imputed datasets. We utilize Rubin's (1987) formulas to use the imputed datasets to account for imputation error. In addition, all estimates are weighted to reflect unequal selection probabilities and response rates. All models are estimated in Stata. We estimate two types of models. The logistic model, which reports estimates of log odds ratios, estimates the probability that a firm that was in business during 2004 went out of business sometime between 2004 and 2008. We then estimate a proportional hazards model including information on credit ratings and geographic controls that vary over the 2004–2008 period. Estimates are presented in Tables 4 and 5. Each table includes estimates for the baseline model and for the baseline model with geographic controls.

6.1 Logistic Model

The variables described above are used to predict the probability of discontinuance using reduced form logistic models (Table 4). The dependent variable is equal to one if the firm in the 2004 sample discontinued sometime between 2004 and 2008. The table includes estimates of the log odds ratio. We estimate the models both with and without the additional geographic controls.[20]

6.1.1 Baseline Model

The baseline model includes credit quality variables, credit access measures, basic firm characteristics, financial ratios,[21] and dummies for Census divisions and major

[20] Additional specifications were estimated with fewer controls, or different combinations of variables. The estimated coefficients and odds ratios are robust to these other specifications. Results from these models can be found in Mach and Wolken (2011).

[21] We include only one financial ratio of each type. The specific ratio chosen from each of the five types of financial ratios was the one that performed best in predicting that the firm was out of business.

Table 4 Logs odds ratios from logistic regression model of going out of business

	Without geographic controls	With geographic controls
Credit quality variables		
Firm credit score	0.994	0.994
	(2.07)**	(2.08)**
Firm or owner declared bankruptcy	1.120	1.118
	(0.27)	(0.27)
Firm or owner delinquent 3+ times	1.302	1.321
	(1.11)	(1.16)
Judgment against the firm or owner	2.496	2.586
	(2.78)***	(2.86)***
Had no D&B rating in 2003	6.338	6.392
	(5.28)***	(5.31)***
Credit access variables		
Firms fearing denial	0.909	0.897
	(0.42)	(0.48)
Borrowing on credit cards	1.437	1.451
	(1.81)*	(1.87)*
Borrowing on trade credit	1.578	1.592
	(2.33)**	(2.39)**
Been denied trade credit	0.865	0.863
	(0.41)	(0.42)
Firms using real estate as collateral	1.201	1.185
	(0.90)	(0.84)
FHFA home prices (% Δ 2004–08)	0.990	0.993
	(1.64)	(0.97)
Firm/owner variables		
Firm age	0.986	0.985
	(1.58)	(1.68)*
Firms employment	1.000	1.000
	(0.00)	(0.06)
Log of firm assets	1.076	1.079
	(1.10)	(1.17)
Log of firm sales	1.025	1.025
	(0.35)	(0.34)
Firm has zero or negative equity	1.055	1.047
	(0.24)	(0.21)
Owner over 60 years old	1.385	1.351
	(1.52)	(1.41)
Owner > median experience	1.016	1.036
	(0.08)	(0.18)
Financial ratio variables		
Liabilities/assets ratio	0.976	0.976
	(1.31)	(1.35)
Cash/sales ratio	1.398	1.401
	(2.76)***	(2.77)***
Accounts payable/sales ratio	0.400	0.402
	(1.71)*	(1.76)*

(continued)

Table 4 (continued)

	Without geographic controls	With geographic controls
Pretax profit/asset ratio	1.002	1.002
	(0.91)	(0.96)
Pretax profit/liabilities ratio	1.000	1.000
	(0.87)	(0.86)
Division variables		
New England	0.607	0.637
	(1.66)*	(1.37)
Mid Atlantic	0.993	0.956
	(0.02)	(0.10)
East N Central	0.592	0.618
	(1.81)*	(1.49)
West N Central	2.103	2.065
	(2.56)**	(2.23)**
South Atlantic	0.513	0.550
	(1.62)	(1.39)
East S Central	1.202	1.235
	(0.70)	(0.73)
Mountain	0.933	1.015
	(0.19)	(0.04)
Pacific	1.079	1.226
	(0.26)	(0.57)
Industry variables		
Construction and mining	1.933	1.975
	(2.35)**	(2.44)**
Manufacturing	1.548	1.549
	(1.40)	(1.40)
Transportation	1.110	1.137
	(0.24)	(0.30)
Wholesale trade	1.311	1.305
	(0.73)	(0.72)
Retail trade	1.565	1.601
	(1.89)*	(1.96)*
Insurance agents and real estate	1.510	1.532
	(1.15)	(1.19)
Professional services	1.120	1.135
	(0.47)	(0.52)
Geographic variables		
Population density (% Δ 2004–08)		1.000
		(0.24)
Average wage (% Δ 2004–08)		1.046
		(2.03)**
Unemployment rate (% Δ 2004–08)		1.003
		(0.54)
Establishment density (% Δ 2004–08)		0.986
		(0.58)

(continued)

Table 4 (continued)

	Without geographic controls	With geographic controls
Per capital income (% Δ 2004–08)		0.969
		(1.59)
Group significance tests		
F Test: Division variables = 0	2.97	1.85
P-Val: Division variables = 0	0.00	0.06
F Test: Quality of credit variables = 0	8.20	8.31
P-Val: Quality of credit variables = 0	0.00	0.00
F Test: Credit access variables = 0	2.54	2.59
P-Val: Credit access variables = 0	0.03	0.02
F Test: Size variables = 0	1.24	1.33
P-Val: Size variables = 0	0.29	0.25
F Test: Ratio variables = 0	2.42	2.43
P-Val: Ratio variables = 0	0.03	0.03
F Test: Industry variables = 0	1.24	1.30
P-Val: Industry variables = 0	0.28	0.24
F Test: Geographic variables = 0		1.08
P-Val: Geographic variables = 0		0.37

Notes: t statistics in parentheses; * significant at 10%; ** significant at 5%; *** significant at 1%. Estimates and standard errors have been adjusted for multiple imputations

SIC industries. Estimates are provided in column 1 of Table 4. Results for joint hypotheses tests of groups of variables (e.g., credit quality variables) are reported at the bottom of Table 4. By groups, we find that the credit quality and access measures are jointly significant at the 1% level, as are the Census division dummy variables. Financial ratios and credit constraint variables were significant at the 5% level. By group, firm size measures – employment, sales, assets, profit and firm age – and industry dummies were not significant.

On an individual basis, several of the credit quality and credit access variables are significant. For each percentage point the firm's credit score (which ranges from 1 to 100) is above the average score, the odds of no longer being in business decreases 0.6% [(.994−1.0)*100]. Being rated by D&B, regardless of the rating, significantly decreases the likelihood of the firm being out of business by 2008. Having judgments in 2004 against the firm or its owner increase the odds of the firm no longer being in business 1.5 times. And borrowing on credit cards and trade credit in 2004 – an indicator of restricted access to credit – increases the odds that a given firm will no longer be in business by 2008 by 44% and 58%, respectively.

The NETS data provide no indication of why a firm is no longer in business. It could be that the firm became unprofitable and had to shut down. However, it could also be the case that the owner simply decided to close the firm, perhaps to retire. We include measures for the owner's age and experience to try to control for this uncertainty. Neither of these variables is marginally significant.

In general, we would expect increasing prices to be associated with increasing access to credit and a negative correlation with firms no longer being in business in 2008. The FHFA index predicts an inverse relationship between firm discontinuance and real estate prices, but is significant at only the 15% level and only in the baseline model without geographic controls.

6.1.2 Geographic Controls

The second column of Table 4 includes additional measures specific to the geographic area where the firm is located. These measures include percentage changes between 2004 and 2008 in population density, average wages, unemployment rate, establishment density, and per capita income. Jointly, these measures are not significant and, again, the other coefficients are robust to their inclusion. Individually, increases in average wages are significantly associated with an increase in the odds of failure by 2008.

6.2 Proportional Hazard Model

The NETS data provide a yearly measure of whether the firm was in operation which allows us to incorporate a sense of timing into the estimates. We estimate a proportional hazard model of firm survival: given that the firm has survived until time t, what is the likelihood that it will fail at time t + 1. We use the same controls that we used in the logistic models, but examine failure in each year. The resultant hazard ratios are presented in Table 5. These models are also estimated using five implicates and weights, as discussed above.

The results are similar to the logistic models; variables that are significant in the logistic models are generally significant in the hazard models and vice versa. As with logistic estimates, the hazard ratios are robust across specifications.

6.2.1 Baseline Model

Results from the baseline model are provided in column 1 of Table 5. By groups, we find that the credit quality variables, credit constraint variables, financial ratios and Census division dummy variables are jointly significant at the 1% level. None of the other groups of variables are jointly significant at traditional levels.

Individually, the credit score is once again significant. A 1% point change in the credit score decreases the instantaneous probability of going out of business by 0.6%. A judgment rendered against the firm or owner roughly doubles the likelihood of going out of business. Not being rated by D&B increases the likelihood of the firm going out of business fourfold in each period. Borrowing on credit cards or trade credit in 2004 – an indicator of restricted access to credit – increases the yearly

Table 5 Hazard ratios from proportional hazard model of time until out of business

	Without geographic controls	With geographic controls
Credit quality variables		
Firm credit score	0.994	0.994
	(2.09)**	(2.06)**
Firm or owner declared bankruptcy	1.034	1.080
	(0.09)	(0.21)
Firm or owner delinquent 3+ times	1.293	1.263
	(1.23)	(1.11)
Judgment against the firm or owner	2.002	2.086
	(2.45)**	(2.55)**
Has a no D&B rating	4.967	5.032
	(5.70)***	(5.67)***
Credit access variables		
Firms fearing denial	0.901	0.904
	(0.49)	(0.47)
Borrowing on credit cards	1.377	1.397
	(1.74)*	(1.83)*
Borrowing on trade credit	1.487	1.502
	(2.19)**	(2.24)**
Been denied trade credit	0.900	0.864
	(0.33)	(0.46)
Firms using real estate as collateral	1.203	1.197
	(0.99)	(0.94)
FHFA home prices (% Δ $year_{t-1}$-$year_t$)	0.999	1.000
	(0.08)	(0.01)
Firm/owner variables		
Firm age	0.987	0.988
	(1.59)	(1.50)
Firms employment	0.999	0.999
	(0.34)	(0.33)
Log of firm assets	1.063	1.071
	(0.97)	(1.08)
Log of firm sales	1.042	1.030
	(0.61)	(0.43)
Firm has zero or negative equity	1.036	1.068
	(0.17)	(0.31)
Owner over 60 years old	1.323	1.325
	(1.43)	(1.44)
Owner > median experience	1.005	1.026
	(0.03)	(0.15)
Financial ratio variables		
Liabilities/assets ratio	0.980	0.964
	(1.17)	(1.51)
Cash/sales ratio	1.339	1.328
	(3.62)***	(3.49)***
Accounts payable/sales ratio	0.465	0.445
	(1.60)	(1.51)

(continued)

Table 5 (continued)

	Without geographic controls	With geographic controls
Pretax profit/asset ratio	1.001	1.001
	(0.84)	(0.96)
Pretax profit/liabilities ratio	1.000	1.000
	(1.00)	(0.91)
Division variables		
New England	0.691	0.680
	(1.32)	(1.36)
Mid Atlantic	0.968	1.053
	(0.08)	(0.13)
East N Central	0.610	0.595
	(1.81)*	(1.78)*
West N Central	1.732	1.764
	(2.11)**	(2.14)**
South Atlantic	0.595	0.512
	(1.34)	(1.66)*
East S Central	1.241	1.122
	(0.91)	(0.47)
Mountain	0.990	0.988
	(0.03)	(0.04)
Pacific	1.091	1.096
	(0.31)	(0.32)
Industry variables		
Construction and mining	1.881	1.882
	(2.48)**	(2.51)**
Manufacturing	1.484	1.487
	(1.40)	(1.40)
Transportation	1.158	1.188
	(0.36)	(0.43)
Wholesale trade	1.374	1.380
	(0.90)	(0.91)
Retail trade	1.488	1.476
	(1.81)*	(1.79)*
Insurance agents and real estate	1.549	1.489
	(1.29)	(1.13)
Professional services	1.111	1.095
	(0.47)	(0.40)
Geographic variables		
Population density (year $_{t-1}$)		1.015
		(0.29)
Average wage (year $_{t-1}$)		1.001
		(0.89)
Unemployment rate (year $_{t-1}$)		0.963
		(0.90)
Establishment density (year $_{t-1}$)		1.000
		(0.43)
Per capital income (year $_{t-1}$)		1.000
		(1.50)

(continued)

Table 5 (continued)

	Without geographic controls	With geographic controls
Group significance tests		
F Test: Division variables = 0	2.40	2.34
P-Val: Division variables = 0	0.01	0.02
F Test: Quality of credit variables = 0	9.13	9.18
P-Val: Quality of credit variables = 0	0.00	0.00
F Test: Credit access variables = 0	2.59	2.65
P-Val: Credit access variables = 0	0.02	0.02
F Test: Size variables = 0	1.29	1.17
P-Val: Size variables = 0	0.26	0.32
F Test: Ratio variables = 0	3.55	3.60
P-Val: Ratio variables = 0	0.00	0.00
F Test: Industry variables = 0	1.30	1.32
P-Val: Industry variables = 0	0.24	0.23
F Test: Geographic variables = 0		0.92
P-Val: Geographic variables = 0		0.47

Notes: t statistics in parentheses; * significant at 10%; ** significant at 5%; *** significant at 1%. Estimates and standard errors have been adjusted for multiple imputations

likelihood that a given firm will go out of business by 38% and 49%, respectively. And as was true with the logit models, owner age, owner experience, and the FHFA housing price index were not significant.

6.2.2 Geographic Controls

The second column of Table 5 includes additional measures specific to the geographic area where the firm is located. These measures include a 1-year lagged value of population density, average wages, unemployment rate, establishment density, and per capita income. Jointly and individually, these measures are not significant and, again, the other coefficients are robust to their inclusion.

7 Conclusions and Further Work

In sum, the results suggest that there are large differences between the continuing and discontinuing firms. Generally, credit access and credit quality measures in 2004 are important indicators of whether the firm will be in business in by 2008. Other significant characteristics include location – being headquartered in the West North Central, East North Central or South Atlantic Census divisions – and industry – being a construction or retail trade business. Equally informative is the finding that generally financial ratios and geographic controls were not significant in predicting whether the firm will be in business in 2008.

It is important to keep in mind that the recent economic turmoil is not representative of all periods. Despite the extraordinary times, the credit access measures do seem to be predictive of firm performance. In future research we would like to address the question of whether some of the credit constrained firms may have fared better had the economy performed better by comparing firm failure during this period to earlier periods.

References

Altman EI (1968) Financial ratios, discriminant analysis and the prediction of corporate bankruptcy. Journal of Finance 23(4):589–609 (September)

Altman EI, Sabato G (2007) Modeling credit risk for SMEs: evidence from the U.S. market. ABACUS 43(3):332–357

Argenti J (1976) Corporate collapse, the causes and symptoms. McGraw Hill, New York

Avery R, Bostic R, Samolyk K (1998) The role of personal wealth in small business finance. J Bank Finance 22:613–673

Beaver W (1966) Financial ratios as predictors of failure. Empirical research in accounting: selected studies. J Account Res 4:71–111 (Supplement)

Blanchflower D, Evans D (2004) The role of credit cards in providing financing for small businesses. Paym Card Econ Rev 2:75–95, Winter

Boden RJ (2000) Analyses of business dissolution by demographic category of business ownership. Final report prepared for the Office of Advocacy of the U.S. Small Business Administration, December (mimeo)

Briggs DH, MacLennan UA (1983) The prediction of private company failure. Eur Manage J 2(1):66–72

Calhoun, CA (1996) OFHEO house price indexes: HPI technical description, working paper. http://www.FHFA.gov/webfiles/896/hpi_tech.pdf

Cavalluzzo K, Wolken J (2005) Small business loan turndowns, personal wealth, and discrimination. J Bus 78(6):2153–2177. 50(11), 1138–1148

Dunne T, Roberts MJ, Samuelson L (1988) Patterns of firm entry and exit in US manufacturing industries. Rand J Econ 19(4):495–515

Hall G (1992) Reasons for insolvency amongst small firms – a review and fresh evidence. Small Bus Econ 4(3):237–250

Headd B (2003) Redefining business success: distinguishing between failure and closure. Small Bus Econ 21:51–61

Holman W, Fletcher H (1989) Assessing the impact of regulation on the business failure rates in industries dominated by small firms. In: Small firm financial research symposium, School of Business Administrative Sciences, California State University, Fresno

Hudson J (1986) An analysis of company liquidations. Appl Econ 18(2):219–235

Hyytinen A, Pajarinen M (2007) Is the cost of debt capital higher for younger firms? Scott J Pol Econ 54(1):55–71

Jovanovic B (1982) Selection and the evolution of industry. Econometrica 50(3):649–670

Keasey K, Watson R (1986) The prediction of small company failure: some behavioural evidence for the UK. Account Bus Res 17:49–57

Keasey K, Watson R (1991) The state of the art of small firm failure prediction: achievements and prognosis. Int Small Bus J 9(4):11–29

Mach TL, Wolken JD (2006) Financial services used by small businesses: evidence from the 2003 survey of small business finances. Fed Reserv Bull 92:167–195

Mach TL, Wolken JD (2011) Examining the impact of credit access on small firm survivability. Board of Governors of the federal reserve system, finance and economics discussion series 2011–35 (August)

Musso P, Schiavo S (2008) The impact of financial constraints on firm survival and growth. J Evol Econ 18:135–149

National Federation of Independent Businesses (2010) Small business economic trends. http://www.nfib.com/tabid/83/Default.aspx

National Opinion Research Center (2005) The 2003 survey of small business finances methodology report. Mimeo. http://www.federalreserve.gov/pubs/oss/oss3/ssbf03/ssbf03home.html

Ohlson J (1980) Financial ratios and the probabilistic prediction of bankruptcy. J Account Res 18 (1):109–131

Peel MJ, Peel DA (1987) Some further empirical evidence on predicting private company failure. Account Bus Res 18(69):57–66

Peterson M, Rajan R (1997) Trade credit: theories and evidence. Rev Financ Stud 10(3):3–37, Autumn

Robb A, Robinson D (2010) The capital structure decisions of new firms. National Bureau of Economic Research, working paper no. 16272, August

Rubin DB (1987) Multiple imputation for nonresponse in surveys. Wiley, New York

Scott R (2009) Plastic capital: credit card debt and new small business survival, working paper, Monmouth University

Shailer G (1986) Experimental failure prediction models for small companies, Discussion paper no. 25, Institute of Industrial Economics, Newcastle

Shailer G (1989) Research note: the predictability of small enterprise failures: evidence and issues. Int Small Bus J 7(4):54–58

Simmons P (1989) Bad luck and fixed costs in personal bankruptcies. Economic Journal 99:92–107

Small Business Administration (2009) Frequently asked questions, Office of Advocacy website. http://web.sba.gov/faqs/faqIndexAll.cfm?areaid=24

Storey D, Keasey K, Watson R, Wynarczk P (1987) The performance of small firms. Croom-Helm, London

Taffler RJ (1982) Forecasting company failure in the UK using discriminant analysis and financial ratio. J Roy Stat Soc 145(3):342–358

Wadwhani B (1986) Inflation, bankruptcy, default premia and the stock market. Econ J 96:120–138

Walls & Associates (2009) National establishment time-series (NETS) Database: 2008 database description. http://www.youreconomy.org/nets/NETS%20Database%20Description2009.pdf

Internal Capital Market and Investment Decisions in Small and Medium-Sized Groups

Donato Iacobucci

1 Introduction

The financial crisis that hit Western economies in 2008 and 2009 had a strong effect on the investment plans of firms as a result of a combined real and financial shock: a sharp decline in product demand and changing conditions in the cost and availability of bank loans (Campello et al. 2010; Ivashina and Scharfstein 2010).

The financing problems were particularly acute for small and medium-sized firms which are more dependent on bank loans for their external financing. The availability of internal liquidity becomes crucial for investment spending when firms face financing constraints (Stein 2003). In fact, empirical evidence shows that small firms are more dependent on the internal generation of funds than larger firms (Fazzari et al. 1988). Not only are large firms less constrained than small firms when raising external funds, but they can also benefit from the possibility to transfer liquidity from cash-producing to cash-absorbing divisions. This is not the case for small firms that operate mostly in just one line of business.

However, not all small firms operate as stand-alone firms as they may be part of a business group: i.e. a set of firms that are owned and controlled by the same entrepreneur(s). The latter may be a single person, members of the same family or a group of people. They can directly own and control several companies (horizontal group) or control them indirectly through a cascade of companies (pyramidal group). The latter is the most widespread form, even in the small business sector; the most common structure is a two-layer pyramid with a head company controlling several others (Cainelli and Iacobucci 2007). One of the features of business groups is the possibility to manage an internal capital market by channeling financial resources from one company to another, not dissimilar from

D. Iacobucci (✉)
Dipartimento di Ingegneria dell'Informazione, Università Politecnica delle Marche, Via Brecce Bianche, Monte Dago 60121, Ancona, Italy
e-mail: d.iacobucci@univpm.it

G. Calcagnini and I. Favaretto (eds.), *Small Businesses in the Aftermath of the Crisis*, Contributions to Economics, DOI 10.1007/978-3-7908-2852-8_11,

what happens in multidivisional companies. Moreover, the legal autonomy of member companies multiplies the units that can autonomously raise external funds.

The last decade witnessed a surge in studies about business groups (Carney et al. 2011). This is the result of an increasing awareness by researchers, policy makers and business people of the widespread presence of business groups and of the diversity of reasons that justify their presence. To control a company it is not necessary to own the 100% of shares: it is enough to hold 50% plus one share. In listed companies the controlling share can be less than 50%, depending on the size of the firm and the dispersion of ownership. This means that in firms belonging to business groups it is common to observe the presence of minority shareholders. The presence of minority (i.e. non-controlling) owners in companies belonging to business groups is the feature that has attracted the attention of researchers interested in explaining the presence of such structures. The prevailing explanation for the formation of business groups is that they are used as a mechanism for separating ownership and control, i.e. to minimize the capital needed to control a business or to maximize the return on capital invested in new businesses. The first mechanism (equity leverage) is observed in large pyramidal groups (Faccio and Lang 2002; Claessens et al. 2000). The second mechanism can also be observed in small and medium-sized groups where minority shareholders can be expropriated to the advantage of the controlling entrepreneur (Almeida and Wolfenzon 2006). There are two basic mechanisms through which expropriation takes place: the first is by setting excessive benefits (such as remunerations or fringe benefits) for the controlling entrepreneur(s); the second is by diverting resources from companies in which the controlling entrepreneur(s) has a lower stake towards companies in which the controlling entrepreneur(s) has a higher stake (Almeida and Wolfenzon 2006). Most often the latter company is the one at the head of the pyramid (Morck and Yeung 2003).

The diversion of resources from one company to another in business groups is referred to as tunneling. It is an aspect that has greatly attracted the attention of researchers (Bertrand et al. 2002; Cheung et al. 2006; Friedman et al. 2003; Gopalan et al. 2007). These studies refer almost exclusively to large groups, often the largest in a country, and to emerging countries. Business groups are supposed to be more widespread in emerging countries as they can help overcome the deficiencies of market institutions (Colpan et al. 2010). However, several studies have demonstrated that business groups are also largely widespread in developed countries and in the small business sector (Rosa and Scott 1999; Cainelli and Iacobucci 2011; Loiseau 2001; Cayssialis et al. 2007; Iacobucci 2002). This evidence has induced researchers to question the traditional financial explanations of business groups. Organizational rather than financial reasons seem to prevail when explaining the presence of groups in the small business sector: business groups can be seen as the organizational form most suited to accommodate growth in related fields and to enlarge the entrepreneurial team (Iacobucci and Rosa 2010; Iacobucci 2002; Lechner and Leyronas 2009). Moreover, in small groups, ownership concentration is quite high and this reduces the importance of the financial mechanisms previously highlighted.

The existence of an internal capital market in business groups can be considered not only as a mechanism for expropriating minority shareholders but also as a mechanism that can have potential benefits for affiliated companies and for the group as a whole.[1] In a recent work Hamelin (2011) demonstrates that, in the case of French small and medium-sized groups, affiliation promotes stability of returns and efficiency rather than expropriation. There is also an increasing amount of literature demonstrating that the group can be beneficial in promoting the R&D investment of member companies and is more efficient in taking advantage of the knowledge spillover arising from innovative efforts (Chang et al. 2006; Belenzon and Berkovitz 2010; Hsieh et al. 2010).

The presence of an internal capital market is specifically relevant when there are difficulties in raising external finance. These difficulties can be associated with the type of investment – intangible assets are more difficult to finance than tangible assets – or with the overall characteristics of the firm asking for external funds: young and small firms have more difficulties than old and large firms in accessing the capital market. The internal capital market of the group can help overcome these difficulties. As a result, firms belonging to a group should be less constrained in their investment policy than stand-alone firms. The legal autonomy of firms affiliated to business groups allows them to directly access capital markets in the same way as their stand-alone counterparts; however, the affiliated firms can also benefit from financial resources provided by other companies in the group. Internal flows of funds in business groups can take different forms: (a) direct equity and credit financing; (b) as a result of the exchanges of goods and services; (c) as a result of non-routine operations such as the buying and selling of assets. A direct measure of the importance and direction of these flows is rather difficult; for this reason most of the literature on this subject relies on indirect measures. One of the ways to assess the presence and relevance of the internal capital market is to observe whether firms belonging to groups are more or less financially constrained than stand-alone firms when financing their investment. Up to now only a few papers have investigated this issue and, like much of the literature on business groups, they refer to large groups (George et al. 2011; Lensink et al. 2003; Shin and Park 1999).

Starting from these premises the main aim of this paper is to analyze whether small and medium-sized firms belonging to groups are more or less financially constrained in their investment policy than their stand-alone counterparts. One of the ways to detect the presence of financing constraints (i.e. difficulties in accessing the capital market) is by measuring the sensitivity of investment to the internal generation of funds (cash flow). Following the seminal paper by Fazzari et al. (1988) a great amount of literature has emerged that has investigated the cash flow-investment relationship in different countries and for different types of firms. Some

[1] The term 'affiliated' or 'member' of a business group is somewhat misleading because belonging to a group is not a 'choice' of the company. The fact that firms belong to a group is the result of established entrepreneurs setting up new companies or acquiring established ones. However, the terms are widely used in the literature and this paper will retain this convention.

contributions have also questioned the possibility of detecting the presence of financing constraints by using investment-cash flow sensitivity (Kaplan and Zingales 1997; Whited 2009). I acknowledge the cautions raised by this literature about the interpretation of the cash flow-investment relationship. However, this paper is less concerned with the causal relationship between cash flow and investment than with the differences in the investment behavior of firms affiliated to business groups and stand-alone firms. In fact, the main aim of this paper is to verify whether there are differences between affiliated and non-affiliated firms in investment-cash flow sensitivity. The focus is on small and medium-sized firms.

The empirical part of the paper is based on the Unicredit dataset. This dataset contains information on a representative sample of Italian manufacturing firms with at least ten employees. We use the X survey that collected information about firms referring to the period 2004–2006. For the same firms, annual reports are available covering a longer period. In the empirical estimates we use a balanced panel covering the period 2001–2006. Up to now this is the first paper to address this issue taking into consideration small and medium-sized firms; i.e. firms with less than 250 employees. The study was not able to extend the analysis to the crisis years of 2008–2009 because of the lack of joint information about financial performance and group affiliation. However, it is likely that the results obtained in the period under analysis will extend to subsequent periods.

Overall the results confirm that firms belonging to groups are less dependent on the internal generation of funds when financing their investments. To the extent that cash flow-investment sensitivity is interpreted as the presence of financial constraints, the results of the paper show that belonging to a business group reduces the constraints when raising funds to finance investments in tangible assets.

The paper is organized as follows. Section 2 examines the literature on business groups, with specific emphasis on those works that analyze the role of internal capital markets and their influence on investment decisions. Section 3 presents the data and the methodology. Section 4 discusses the empirical results. Section 5 presents the main conclusions and highlights the possible extensions of the work.

2 Literature Background and Research Hypotheses

A business group is a set of companies that are legally distinct but belonging to the same owner(s). The prevailing explanation for the presence of business groups is that they are used by entrepreneurs as a financial device to minimize the capital needed to control a business or to maximize the return on the capital invested. This is done by expropriating resources to the minority shareholders of the companies controlled by the group (Almeida and Wolfenzon 2006; Morck and Yeung 2003). The literature on business groups is overwhelming and it is out of the scope of this paper to provide even a brief synthesis. Most of this literature refers to large groups and to emerging economies (Colpan et al. 2010). This overlooks the fact that business groups are largely widespread in developed countries and also in the small business sector (Iacobucci 2002; Carney et al. 2011). There is an increasing awareness by researchers looking at small groups in developed countries that business groups are set up for efficiency reasons rather than as a mere financial device (Iacobucci and Rosa 2005; Lechner and Leyronas 2009; Hamelin 2011). They can be a mechanism for promoting growth and diversification by favoring the formation of entrepreneurial teams (Iacobucci and Rosa 2010), or can raise the innovative effort of member firms by favoring the financing of R&D projects and the exploitation of their results (Chang et al. 2006; Belenzon and Berkovitz 2010; Blanchard et al. 2005). One of the features that distinguishes a business group from a network of independent firms, is that a group can more easily exploit financial synergies. This can be done not only through market exchanges but also in more direct ways, such as the provision of equity capital and credit financing (Gopalan et al. 2007).

Whatever the reasons are for its formation, one of the interesting features of a business group is that it can function as an internal capital market for affiliated firms. The transfer of financial resources from one company to another can be done using several mechanisms. These mechanisms can be grouped in three main areas: (1) equity of credit financing; (2) market-based transactions; (3) non-routine operations. Equity financing can go in both directions: the head of the group may provide additional equity to member firms or the latter may pay more dividends to the controlling firm. Companies can also open direct credit or debit positions with other companies. Both forms are more likely between the head of the group and the controlled firms rather than between controlled firms. When there are input–output exchanges between companies it is possible to use them to transfer resources from one company to another. This could be done by altering the prices at which market exchanges occur. A lighter form of tunneling associated with market exchanges could be that of lengthening or reducing the terms of payments associated with market exchanges. Finally, the transfer of resources between companies in a group can be the result of non-routine operations, such as the buying and selling of assets between the companies in the group.

Most of the empirical studies about internal capital markets in business groups are interested in verifying how and to what extent the controlling owner will use this power to expropriate minority shareholders in controlled companies (Bertrand et al. 2002; Cheung et al. 2006). However, a recent work by Hamelin (2011) has questioned this assumption by demonstrating that, in the case of small groups, the flow of capital can go in both directions, from and to the head of the group, in which the controlling entrepreneur is supposed to retain the largest stake. Hamelin (2011) also suggests that the direction of flows is different according to the business cycle: when member firms are diversely affected by the business cycle, companies less hit by negative shocks can subsidize other companies and vice-versa. The latter mechanism is referred to as propping (Friedman et al. 2003; Dow and Mcguire 2009).

With a few exceptions (Gopalan et al. 2007) the presence of tunneling and propping is detected in an indirect way, by observing the financial performance of member firms. As noted by George and Kabir (2008) this is because most of the transactions between affiliated companies are not disclosed in public reports. For this reason, it is difficult to detect in a direct way the amount of resources transferred between the companies in a group.

One of the main functions of an internal capital market is to provide financial resources to firms that need to make investments and that could be financially constrained when accessing the capital market. For this reason, firms belonging to a group are expected to be less financially constrained in their investment policy than stand-alone firms in similar circumstances. In general, external financing is more expensive than internal financing, as a result of information asymmetries between investors and financiers. Moreover, in some cases external financing may not be available at all. For this reason, the firms that find it most difficult to access external funds are supposed to finance their investments mainly from their internally generated resources (cash flow). In large companies the importance of the internally generated funds derives from the agency problems between managers and shareholders (Myers 2000; Myers and Majluf 1984). In small firms the concentration of ownership is quite high and the owners are directly involved in the management of the company. This attenuates or removes the agency problems between owners and managers; however, the raising of additional outside equity is often ruled out by the desire of the controlling entrepreneur not to dilute his/her controlling share. In Italy, as well as in other European countries, most of the external funds raised by small firms take the form of banks loans. Credit markets are characterized by information asymmetries between providers and users of funds.

For these reasons small firms are expected to rely more on the generation of internal cash flow.

Belonging to a group can mitigate this problem if groups can help member firms by providing funds directly or by facilitating access to external financing; for example, the head of the group can provide collaterals to member firms to facilitate access to bank loans. Firms belonging to groups are normally eager to declare that they belong to a larger structure. This enhances their general market standing and also their credit worthiness.

Up to now only a few studies have investigated whether belonging to a group mitigates the dependence on internal cash flow when financing investment.

Shin and Park (1999) compare the investment–cash flow sensitivity of Korean chaebol and non-chaebol firms. They consider a sample of publicly traded companies and split it into 123 firms belonging to the top 30 chaebols and 194 non-chaebol firms. All are large firms. Shin and Park (1999) show that investment–cash flow sensitivity is low and insignificant for chaebol firms but is high and significant for non-chaebol firms; moreover, the chaebol firms' investment is significantly affected by the cash flow of other firms within the same chaebol. This empirical evidence is interpreted by the authors as evidence of the presence of an internal capital market in chaebols that reduces the financing constraints of affiliated firms.

Lensink et al. (2003) compare the investment behavior of business group affiliates with that of stand-alone companies and test the hypothesis that firms belonging to groups are less financially constrained than stand-alone firms.

They use data about 694 listed Indian companies for the period 1989–1997. They find that the investment-cash flow sensitivity is much lower for group affiliates. This leads the authors to suggest that firms belonging to a business group have better access to external funds, either from within the group or from outside the group, than stand-alone firms.

George et al. (2011) investigate the investment-cash flow sensitivity of a sample of 339 firms listed on the Bombay Stock Exchange, 198 of which belong to a group. Unlike the previously mentioned studies, George et al. (2011) find a strong investment-cash flow sensitivity for both stand-alone and group-affiliated companies and no significant differences in the investment cash-flow sensitivity between the two groups.

All these studies refer to large groups and listed companies. As such, they have enough market standing by themselves when raising external funds; belonging or not to a group is not as relevant as it could be for smaller, non-listed firms.

This paper investigates the investment-cash flow sensitivity in the small firm sector by comparing firms belonging and not belonging to business groups. Our hypothesis is that firms belonging to business groups are expected to be less financially constrained than stand-alone firms; i.e. their investments are less sensitive to the generation of internal cash flow thanks to the possibility to rely on an internal capital market and to the advantages of belonging to a group when accessing external financing.

3 Data and Methodology

To test the above mentioned hypothesis we use data from the Unicredit (formerly Capitalia) dataset (Unicredit 2008). The dataset is based on a periodical survey of a representative sample of Italian manufacturing firms with more than ten employees. We use the data from the X survey, referring to 2006.

The survey asks interviewed companies whether they belong to a group and their position within the group, i.e. whether the firm is at the head of the group, in an intermediate position or at the bottom of the pyramid. Further information allows a detailed analysis of the ownership structure of the firms, since the dataset contains details on both ownership shares and the type of owners.

The dataset contains information on 5,075 firms, of which 979 belong to a business group. Of the latter, 226 are at the head of the group and the remaining are controlled companies. We focus on small firms, i.e. on those firms with a number of employees lower than 250. The firms with less than 250 employees are 4,611, of which 746 belong to a group. The original dataset has been integrated with data from company annual reports spanning from 1998 to 2006. Balance sheet data are not available for all the firms and all years; the valuable cases decline steadily from 4,445 in 2006 to 2,975 in 1998. For this reason in econometric estimates we check the results by considering different periods. In order to retain a balanced panel with a sufficient number of cases we restrict the period of observation by considering the years 2001–2006.[2]

To test the investment-cash flow sensitivity we use a standard Euler investment equation with the addition of a cash flow variable (Goergen and Renneboog 2001):

$$\left(\frac{I}{K}\right)_{i,t} = \beta_0 + \beta_1\left(\frac{I}{K}\right)_{i,t-1} + \beta_2\left(\frac{I}{K}\right)^2_{i,t-1} + \beta_3\left(\frac{S}{K}\right)_{i,t-1} + \beta_4\left(\frac{D}{K}\right)_{i,t-1} + \beta_5\left(\frac{CF}{K}\right)_{i,t-1} + \gamma_t + \varphi_i + \varepsilon_{i,t} \quad (1)$$

where:

I = investment in fixed assets
K = fixed assets measured at the beginning of the period
S = sales
CF = cash flow
γ = time specific effects
φ = firm specific effects

Most studies on the sensitivity of investment to cash flow use Tobin's Q, instead of the previous year sales, as a variable capturing the opportunities for investment. This paper uses the Euler model of investment instead of Q as only 39 out of the 4,611 companies with less than 250 employees are listed on the stock exchange.

There are other variations in the way of formulating the investment-cash flow relationships. Some authors use the stock of cash and securities at the beginning of the year rather than the cash flow in the previous year. In the accelerator model the ratio of sales to capital is substituted with the variation in sales.

An interesting variation of the Euler investment model is proposed by Goergen and Renneboog (2001); multiplying both terms of Eq. 1 by K/S they express all the variables in terms of unit of sales rather than capital and remove from the independent variables the term S/K. This transformation has the advantage of more stable value over time and of eliminating the problem of anomalous values resulting from very low values of K. Another way of dealing with the latter problem is the one proposed by Cleary (1999, p. 678) of 'winsoring' the data by imposing minimum and maximum cut-off values for all the parameters: 2 and −2 for the investment-capital ratio, 5 and −5 for the cash flow-capital ratio, etc. Whatever the way of 'winsoring' the data, the imposed cut-off values are always arbitrary. In the following statistics and estimates we used the cut-off proposed by Cleary (1999).

[2] A possible limitation of the analysis is that during the period under analysis some of the firms could have altered their position as an affiliated or non-affiliated firm. Previous empirical studies on small groups (Iacobucci 2002) demonstrate that groups are rather stable over time and that changes in status should regard a tiny minority of firms. However, as a robustness check for this issue we replicated the analysis for periods of different lengths, obtaining the same results.

Table 1 List of variables used in the estimates

Variable	Way of calculation
Investment rate	Investment in tangible assets at time *t* on the value of tangible assets at the beginning of the period = I_{it}/K_{it}
Cash flow	Net income + depreciation and amortization at time *t* on tangible assets at the beginning of the period = CF_{it}/K_{it}
Sales to capital ratio	Sales at time *t* on tangible assets at the beginning of the period = S_{it}/K_{it}
Leverage	Financial debt on tangible assets = FD_{it}/K_{it}
Liquidity	Liquidity (cash + marketable securities) on tangible assets = L_{it}/K_{it}
Group	Dummy taking the value of 1 if the firm belongs to a group, 0 otherwise
Head	Dummy taking the value of 1 if the firm belongs to a group and is at the head of the group
Controlled	Dummy taking the value of 1 if the firm belongs to a group and is controlled by other companies
Age	Age of the firm from set-up, in years
Size	Size of the firm, measured as the log of employees

We also tried other cut-off values; they influence the means of the variables but do not alter the overall results of the analysis.

There are two ways of evaluating if and how the cash flow-investment relationship differs in different groups of firms (in our case group-affiliated and stand-alone firms): the first way is that of splitting the sample into different groups and comparing the coefficients of the cash flow variable (Audretsch and Elston 2002); the second consists in introducing interactive dummies among the independent variables (George et al. 2011; Lensink et al. 2003). This study makes use of both techniques.

Table 1 lists the variables that will be used in the empirical analysis and the way they are calculated. Flow variables (such as investment, cash flow, sales) are standardized on the capital at the beginning of the year; stock variables (such as debt, liquidity, etc.) are standardized with the capital measured at the same point in time.

Table 2 summarizes descriptive statistics for the main variables, by splitting the sample into firms belonging and not belonging to groups.

The main differences between the two groups are in the average size of the firms and in the leverage, that are both higher in affiliated firms than in stand-alone ones. In the empirical estimates we also use sectorial dummies based on the Pavitt (1984) classification.[3]

[3] The Pavitt classification of sectors based on innovation patterns (Pavitt 1984). The classification distinguishes between four macro sectors: scale-intensive, supplier-dominated, science-based and specialized suppliers. The supplier-dominated sector is used as reference.

Table 2 Summary statistics for the pooled sample

Variables	Total firms		Non-group firms		Group firms	
	Mean	Std.dev.	Mean	Std.dev.	Mean	Std.dev.
Investment rate	0.182	0.594	0.193	0.601	0.123	0.545
Sales to capital ratio	7.178	5.144	7.141	5.127	7.369	5.226
Cash flow to capital ratio	0.501	0.968	0.487	0.929	0.574	1.145
Leverage	1.090	1.594	1.048	1.579	1.305	1.651
Age	42.5	41.8	37.0	35.4	70.6	57.6
Size (employees)	19.3	22.4	19.2	22.7	20.1	21.3

Source: elaborations on Unicredit dataset

4 Results

Table 3 presents the correlation matrix between the main variables, calculated on the pooled sample. Some of the variables present high values of linear correlation; however, the mean of the variance inflation factor is 1.2 and the condition number is 10.9, which excludes the presence of multicollinearity problems.

Table 4 presents the results of the regressions on the pooled sample based on random effects estimators and using cluster robust standard errors. Model 3 tests the main hypothesis by introducing an interactive dummy as shown in Eq. 2

$$\left(\frac{I}{K}\right)_{i,t} = \alpha + \beta_0 + \beta_1\left(\frac{I}{K}\right)_{i,t-1} + \beta_2\left(\frac{I}{K}\right)^2_{i,t-1} + \beta_3\left(\frac{S}{K}\right)_{i,t-1} + \beta_4\left(\frac{D}{K}\right)_{i,t-1} + \beta_5\left(\frac{CF}{K}\right)_{i,t-1} + \beta_6\left(\frac{CF}{K}\right)_{i,t-1} * (1 - DG) + \mu_{i,t} + \varepsilon_{i,t} \qquad (2)$$

All the symbols have the same meaning as in Eq. 1 and DG is a dummy variable taking the value of 1 if the firm belongs to a group and 0 otherwise. In this specification, the coefficient β_5 measures the cash flow sensitivity of companies affiliated to groups and $\beta_5 + \beta_6$ measures the cash flow sensitivity of stand-alone firms (Lensink et al. 2003, p. 107).

The coefficients of the standard Euler equation show the expected signs and are robust to the introduction of firms and industry dummies (models 1 and 2). For the pooled sample, the cash flow variable is positively associated with the rate of investment of companies.[4] Model 3 shows that when introducing the group

[4] We use the term association because our aim is not that of a causal analysis. Moreover, we are not interested in discussing to what extent a positive association between cash flow and investment can be considered as a measure of financing constraint. Our aim is only that of assessing the presence of differences in the cash flow-investment sensitivity between affiliated and non-affiliated companies.

Table 3 Correlation matrix for the pooled sample

	Investment rate	Cash flow	Sales to capital ratio	Leverage	Age	Size
Investment rate	1					
Cash flow	0.2039	1				
Sales to capital ratio	0.2429	0.4519	1			
Leverage	−0.1267	0.0595	0.3174	1		
Age	−0.0839	−0.0821	−0.1602	0.0386	1	
Size	−0.0904	−0.0245	−0.0793	0.1731	0.2507	1

Table 4 Regression results of panel estimations with random effects and cluster robust standard errors

	Model 1	Model 2	Model 3	Model 4
Investment (−1)	−0.2075***	−0.2063***	−0.1613***	−0.2054***
	(−12.31)	(−12.29)	(−9.87)	(−12.28)
Investment (−1) sq.	0.0605***	0.0574***	0.0461***	0.0575***
	(6.40)	(6.08)	(5.02)	(6.10)
Sales to capital ratio (−1)	0.0205***	0.0189***	0.0166***	0.0188***
	(20.14)	(18.35)	(15.93)	(18.32)
Cash flow (−1)	0.0281***	0.0306***	−0.0044	0.0284***
	(4.44)	(4.86)	(−0.62)	(4.44)
Leverage (−1)	−0.0200***	−0.1334***	−0.0124***	−0.0134***
	(−6.52)	(−4.23)	(−3.88)	(−4.25)
Cash flow * (1 – group dummy)			0.0868***	
			(9.45)	
Group dummy				−0.0674***
				(−6.05)
Cash flow * group dummy				0.213
				(1.21)
Size		−0.0448	−0.0403***	−0.0380***
		(−9.93)	(−9.01)	(−8.13)
Age		−0.0134	−0.0126***	−0.0145***
		(−4.16)	(−3.96)	(−4.51)
Scale-intensive		0.0089	0.0087	0.0114
		(0.94)	(0.92)	(1.21)
Specialized		−0.0212	−0.0200***	−0.0183**
		(−2.47)	(−2.32)	(−2.13)
Science-based		−0.0661	−0.0734***	−0.0635***
		(−3.58)	(−3.98)	(−3.45)
Year dummies	yes	Yes	Yes	Yes
Observations	20,614	20,614	20,614	20,614
N. of groups	4,338	4,338	4,338	4,338

The regressions include a constant
** = significant at 5%
*** = significant at 1%

dummy, the cash flow coefficient β_5, which refers to companies belonging to a group loses its statistical significance while the coefficient of the interactive dummy is positive and highly significant. According to the previously discussed interpretation, this means that in companies belonging to groups we do not observe investment-cash flow sensitivity, while this sensitivity remains for the stand-alone companies. This result is confirmed in model 4 where the standard Euler model is modified by introducing a group dummy and an interactive term between the group dummy and the cash flow. The coefficient of the latter variable, although positive, is not statistically significant.

The coefficients of the size and age of firms are both negative and statistically significant. This means that the investment rate in tangible assets shows a smoothing pattern with the increase in the size and age of the firm. This is also explained by the fact that with size and age there is an increasing component of investment in financial and intangible assets. The industry dummies show the expected signs: negative in the case of firms belonging to specialized suppliers and science-based firms as they invest more in intangible assets compared with firms in traditional sectors (that are taken as reference); positive, but not statistically significant, in the case of scale-intensive sectors.

The Hausman test obtained by comparing the estimates of Model 2 using random and fixed effects reveals that the preferred model to obtain consistent estimates of the coefficients is the fixed effects model rather than the random effects model. It is worthwhile underlining that the present analysis is not intended to measure the degree of financial constraints in investment decisions but to detect the presence of differences in the cash flow-investment correlations between different groups. To that end, the results of the estimates based on random effects maintain their informative validity, even if the coefficients are not consistent (Cameron and Trivedi 2005, p. 715). However, to obtain more consistent estimates of the cash flow coefficients for affiliated and non-affiliated firms two alternative strategies have been used. The first is to estimate the fixed effects model by splitting the sample into two clusters: firms belonging to groups and firms not belonging to groups; the second strategy is to use the first differences of the variables used in the regression models.

Table 5 shows the results of the estimates obtained using the fixed effects model applied to firms belonging and not belonging to groups. The cash flow coefficient is positive and significant in firms not belonging to business groups; it is still positive but not statistically significant in firms belonging to business groups, thus confirming that for the latter firms the rate of investment is less sensitive to the availability of internal funds.

The other common way of obtaining consistent estimates of coefficients in the presence of fixed effects is that of using the first differences of the variables. With this model it is also possible to retain the use of the interactive variable, as introduced in the specification Eq. 2. The results of the estimates obtained using the first differences of variables are shown in Table 6.

Table 5 Estimates results using the fixed effects model (t values in parenthesis calculated with robust standard errors)

	1 Firms belonging to groups	2 Firms not belonging to groups
Investment (−1)	−0.2486***	−0.4147***
	(−5.00)	(−20.81)
Investment (−1) sq.	0.0026	0.0396***
	(0.09)	(3.65)
Sales to capital ratio (−1)	0.0636***	0.0805***
	(9.29)	(31.53)
Cash flow (−1)	0.0190	0.0610***
	(1.14)	(5.51)
Leverage (−1)	0.1203***	0.1341***
	(4.98)	(10.90)
Year dummies	Yes	Yes
Observations	3,421	17,193
N. of groups	684	3,654
F of regression	36.2	405.9
	Prob. 0.00	Prob. 0.00
Rho	.497	.538

The regressions include a constant
*** = significant at 1%

Table 6 Estimates results of OLS using the first differences of the variables (t values parenthesis calculated on robust standard errors)

	Model 1		Model 2
	Firms belonging to groups	Firms not belonging to groups	
D.Investment (−1)	−0.6107***	−0.7457***	−0.7026***
	(−16.40)	(−45.35)	(−48.24)
D.Investment (−1) sq.	0.0248	0.0497***	0.0387***
	(1.22)	(5.78)	(5.04)
D.Sales to capital ratio	0.0808***	0.1019***	0.0957***
	(13.21)	(39.83)	(41.05)
D.Cash flow	0.0406	0.0551	0.0358
	(2.77)	(5.50)	(4.27)
D.Leverage	0.2447***	0.2262***	0.2177***
	(9.66)	(19.31)	(21.25)
Cash flow*(1 − DG)			0.1247***
			(15.24)
Year dummies	Yes	Yes	Yes
Observations	3,174	15,520	18,694
R sq	0.379	0.467	.464
F	125.74	1141.97	1171.79
	Prob. 0.00	Prob. 0.00	Prob. 0.00

D. = first differences
The regressions include a constant
*** = significant at 1%

Unlike the previous estimates, the cash flow coefficient is positive and significant for both affiliated and non-affiliated firms. However, the sensitivity to cash flow is higher in the case of stand-alone firms than for firms belonging to groups.

Several robustness checks were performed by varying the sample and the models adopted. We performed the empirical analysis by considering different cut-offs for censoring outliers and different time periods. This will change the values of the coefficients but does not alter the results previously discussed.

5 Conclusions

Business groups are largely widespread in developed countries and also in the small business sector. They can be considered an efficient organizational form that entrepreneurs use to grow and diversify the business under their control. Business groups are preferred to other organizational forms because the legal autonomy of member companies allows entrepreneurs to implement the desired mix of control and autonomy in developing and managing a portfolio of businesses. In small and medium-sized groups, those businesses are often closely related to one another, thus allowing the controlling entrepreneur to exploit technological and marketing synergies between controlled companies.

One of the ways in which these synergies can be exploited is by managing in a unitary way the cash flow generated within the group, thus creating an internal capital market. This means the possibility of tunneling financial resources from cash-producing to cash-absorbing companies, thus smoothing the impact of external shocks and taking advantage of firms that are at different stages in the development of their business. The internal capital market also offers the advantage of reducing the need to raise funds from external sources to finance investment. The latter feature is particularly relevant when firms experience difficulties in raising external funds because of information asymmetries in capital markets or a sudden worsening of cost and availability conditions, such as the situation experienced in the financial crisis of 2008–2009. The main hypothesis of this paper is that firms belonging to business groups are expected to be less financially constrained in their investment policies because they can rely on the transfer of resources from other affiliated companies. The presence of an internal capital market in small and medium-sized groups is tested by comparing the cash flow-investment sensitivity between affiliated and non-affiliated companies. This hypothesis has already been empirically tested. However, up to now empirical studies have referred to large groups in emerging economies; to the best of our knowledge this is the first analysis referring to small and medium-sized groups in a developed country.

Overall the results confirm the main hypothesis. Depending on model specification, the internal generation of funds is not significant in explaining the rate of investment in business groups or why its relevance is lower than in stand-alone firms. To the extent that cash flow-investment sensitivity is interpreted as the presence of financial constraints, the results of the paper show that belonging to a business group reduces the financing constraints when raising funds to finance investment.

This analysis could be further developed in several ways. The most obvious would be the temporal extension to include the crisis period of 2008–2009. The sudden worsening in the cost and availability of bank loans should have hit stand-alone firms more than firms belonging to business groups. A further extension of this work could be the consideration of different types of investment. The investment rate considered in this paper refers to tangible assets. However, even in small firms investments in intangible assets, such as R&D, marketing and human resources, are gaining importance. For these investments the problems in raising external funds are higher than in the case of tangible assets. The presence of an internal capital market in business groups, and the resulting reduction in information asymmetries could be specifically beneficial when financing investment in intangible assets.

Business groups are by no means homogeneous in their internal organization; they differ in the degree of diversification, the concentration of activities in their member companies, the degree of centralization or autonomy in key management functions (Cainelli and Iacobucci 2011). This is true for large as well as for small groups. Further analysis should be focused on assessing to what extent these differences influence the efficiency of this organizational structure and how the internal organization of groups is related to the functioning of the internal capital market.

References

Almeida H, Wolfenzon D (2006) A theory of pyramidal ownership and family business groups. Journal of Finance LXI:2637–2680

Audretsch DB, Elston JA (2002) Does firm size matter? Evidence on the impact of liquidity constraints on firm investment behavior in Germany. Int J Ind Organ 20:1

Belenzon S, Berkovitz T (2010) Innovation in business groups. Managt Sci 56:519–535

Bertrand M, Mehta P, Mullainathan S (2002) Ferreting out tunneling: an application to Indian business groups. Q J Econ 117:121–148

Blanchard P, Huiban JP, Sevestre P (2005) R&D and productivity in corporate groups: an empirical investigation using a panel of French firms. Annales d'économie et de statistique 79/80:461–485

Cainelli G, Iacobucci D (2007) Agglomeration, technology and business groups. Edward Elgar, Cheltenham

Cainelli G, Iacobucci D (2011) Business groups and the boundaries of the firm. Manage Decis 49:1549–1573

Cameron AC, Trivedi PK (2005) Microeconometrics: methods and applications. Cambridge University Press, Cambridge, p 100

Campello M, Graham JR, Harvey CR (2010) The real effects of financial constraints: evidence from a financial crisis. J Financ Econ 97:470–487

Carney M, Gedajlovic ER, Heugens PPMAR, Van Essen M, Van Oosterhout J (2011) Business group affiliation, performance, context, and strategy: a meta-analysis. Acad Manage J 54:437–460

Cayssialis J-L, Kremp E, Peter C (2007) Dix Années De Dynamique Financière Des PME En France, Paris, Banque de France Bulletin, p 165
Chang S-J, Chung CN, Mahmood IP (2006) When and how does business group affiliation promote firm innovation? A tale of two emerging economies. Organ Sci 17:637–656
Cheung Y-L, Rau P, Stouraitis A (2006) Tunneling, propping, and expropriation: evidence from connected party transactions in Hong Kong. J Financ Econ 82:343–386
Claessens S, Djankov S, Lang LH (2000) The separation of ownership and control in East Asian corporations. J Financ Econ 58:81–112
Cleary S (1999) The relationship between firm investment and financial status. J Finance 54:673–692
Colpan AM, Hikino T, Lincoln JR (2010) The oxford handbook of business groups. Oxford University Press, Oxford
Dow S, Mcguire J (2009) Propping and tunneling: empirical evidence from Japanese keiretsu. J Bank Finance 33:1817–1828
Faccio M, Lang LH (2002) The ultimate ownership of Western European corporations. J Financ Econ 65:365–395
Fazzari SM, Hubbard RG, Petersen BC (1988) Financing constraints and corporate investment. Brook Pap Econ Act 1:141–195
Friedman E, Johnson S, Mitton T (2003) Propping and tunneling. J Comp Econ 31:732–750
George R, Kabir R (2008) Business groups and profit redistribution: a boon or bane for firms? J Bus Res 61:1004–1014
George R, Kabir R, Qian J (2011) Investment-cash flow sensitivity and financing constraints: new evidence from Indian business group firms. J Multinatl Financ Manage 21:69–88
Goergen M, Renneboog L (2001) Investment policy, internal financing and ownership concentration in the UK. J Corp Finance 7:257–284
Gopalan R, Nanda V, Seru A (2007) Affiliated firms and financial support: evidence from Indian business groups. J Financ Econ 86:759–795
Hamelin A (2011) Small business groups enhance performance and promote stability, not expropriation. Evidence from French SMEs. J Bank Finance 35:613–626
Hsieh T-J, Yeh R-S, Chen Y-J (2010) Business group characteristics and affiliated firm innovation: the case of Taiwan. Ind Market Manage 39:560–570
Iacobucci D (2002) Explaining business groups started by habitual entrepreneurs in the Italian manufacturing sector. Entrepreneurship Reg Dev 14:31–48
Iacobucci D, Rosa P (2005) Growth, diversification and business group formation in entrepreneurial firms. Small Bus Econ 25:65–82
Iacobucci D, Rosa P (2010) The growth of business groups by habitual entrepreneurs: the role of entrepreneurial teams. Entrepreneurship Theory Pract 34:351–377
Ivashina V, Scharfstein D (2010) Bank lending during the financial crisis of 2008. J Financ Econ 97:319–338
Kaplan SN, Zingales L (1997) Do financing constraints explain why investment is correlated with cash flow? Q J Econ 109:565–592
Lechner C, Leyronas C (2009) Small-business group formation as an entrepreneurial development model. Entrepreneurship Theory Pract 33:645–667
Lensink R, van der Molen R, Gangopadhyay S (2003) Business groups, financing constraints and investment: the case of India. J Dev Stud 40:93–119
Loiseau H (2001) Des Groupes De La Taille D'une PME. Un Phenomen En Plein Essor. INSEE Première
Morck R, Yeung B (2003) Agency problems in large family business groups. Entrepreneurship Theory Pract 27:367–382
Myers SC (2000) Outside equity. J Finance 55:1005–1037
Myers SC, Majluf NS (1984) Corporate financing and investment decisions when firsm have information investors do not have. J Financ Econ 13:187–222

Pavitt K (1984) Sectoral patterns of technical change: towards a taxonomy and a theory. Res Policy 13:343–373

Rosa P, Scott M (1999) The prevalence of multiple owners and directors in the SME sector: implications for our understanding of start-up and growth. Entrepreneurship Reg Dev 11:21–38

Shin H-H, Park YS (1999) Financing constraints and internal capital markets: evidence from Korean "chaebols". J Corp Finance 5:169–191

Stein JC (2003) Agency, information and corporate investment. In: Constantinides GM (ed) Corporate finance. Elsevier, Amsterdam, pp 111–165

Unicredit (2008) Decima Indagine sulle imprese manifatturiere Italiane. Milano

Whited TM (2009) What can cash shortfalls and windfalls tell us about finance constraints? In: Calcagnini G, Saltari E (eds) The economics of imperfect markets. Physica-Verlag Berlin Heiderlberg, pp 17–32 see: http://books.google.it/books?id=UsSeCua0Ca0C &printsec=frontcover&hl=it&source=gbs_ge_summary_r&cad=0#v=onepage&q&f=false

Guarantees and Bank Loan Interest Rates in Italian Small-Sized Firms

Giorgio Calcagnini, Fabio Farabullini, and Germana Giombini

1 Introduction

JEL classification: E43, G21, D82

The importance of guarantees in bank lending activity is widely acknowledged, and their role is even recognized in the Basel Capital Accords that foresee a specific regulation for secured loans. Moreover, there is extensive literature that deals with the role of guarantees in determining the cost and the availability of credit, and borrower characteristics that have a greater effect on collateral requirements. In the presence of informational opacity, collateral and guarantees are powerful tools useful for mitigating adverse selection problems that may arise at the loan origination, and moral hazard risk that arises after credit has been granted (Berger and Udell 1990).

This paper aims at analyzing the role played by collateral and personal guarantees on bank-loan interest rates granted to Italian small-sized firms by means of a large dataset drawn from the Central Credit Register for the fourth quarter of the years 2006–2009. The Central Credit Register is an information system regarding the debt of the customers of the banks and financial companies managed by the Bank of Italy.

The opinions presented are solely those of the authors and do not necessarily reflect the positions and views of the Bank of Italy.

G. Calcagnini (✉) • G. Giombini
Dipartimento di Economia, Società, Politica, Università di Urbino "Carlo Bo", via Saffi 42, 61029 Urbino, Italy
e-mail: giorgio.calcagnini@uniurb.it; germana.giombini@uniurb.it

F. Farabullini
Bank of Italy, Rome, Italy
e-mail: fabio.farabullini@bancaditalia.it

G. Calcagnini and I. Favaretto (eds.), *Small Businesses in the Aftermath of the Crisis*, Contributions to Economics, DOI 10.1007/978-3-7908-2852-8_12,

By means of the Central Credit Register the Bank of Italy provides intermediaries with a service intended to improve the quality of the lending of the credit system and ultimately to enhance its stability. The purpose of this paper is twofold: 1) to model and estimate a bank-loan supply function in order to understand the role and the relative weights of the two types of guarantees in the setting of bank interest rates; 2) to understand if and how Italian bank behaviour has changed during the recent economic and financial crisis.[1]

Pozzolo (2004) studied the effects of guarantees on interest rates. In his work he controlled for the presence of guarantees by means of two dummy variables (one for collateral and one for personal guarantees). The novelty of our paper is: (a) the use of a larger database than Pozzolo's, spanning from 2006 to 2009; (b) the focus on producer households companies for which the provision of collateral and personal guarantees has always been the facilitation of bank credit access, especially during economic crises (Cowling 2010)[2]; (c) the use of the ratios of the guarantee value to loan as explanatory variables. Indeed, we expect that interest rates depend upon the relative value of guarantees with respect to the size of the granted loan, and not only upon the presence of guarantees; (d) we deal with a single-nested panel in which firms may be grouped by banks, and estimate "mixed effects" models in which fixed-effects are analogous to standard regression coefficients and are estimated directly. The random effects are not directly estimated but are summarized according to their estimated variances and covariances. These models allow to capture heterogeneity between banks in loan interest rate and how the the impact of personal guarantees and collateral on the loan interest rate varies across banks.

The distinction between collateral and personal guarantees, and their potential different role in bank-borrower relationships, plays an important role in models of bank interest rates. In an asymmetric information context, guarantees play a role in solving different problems that may arise at loan origination (hidden information-adverse selection problems) or after the loan has been granted (hidden action-moral hazard problems).

Furthermore, a principal role is played by the distinction between inside collateral and outside collateral. The former are physical assets owned by the borrower, and is mainly used to order creditors priority in case of borrower's default. The latter are assets posted by third party: in case of default, outside collateral enhances the claim of a single creditor on additional assets external to the debtor. Therefore,

[1] Panetta and Signoretti (2010) show that, during recent economic and financial turmoil, the decrease of Italian bank loans has been affected both by demand and supply factors. As for firms, they show that loan demand declined following the investment contraction, while lower levels of bank loan supply were the joint result of the increase in borrowers' risk and in the degree of bank risk aversion.

[2] Producer households are partnerships or sole proprietorships with a number of employees less than or equal to five.

outside collateral should be more powerful than inside collateral in solving asymmetric information problems.

Personal guarantees are on the other hand contractual obligations of a third party, and they act as external collateral. However, they do not give the lender a specific claim on particular assets, and restrict the actions (s)he could take in case of borrowers' bankruptcy (Berger and Udell 2005).

The Central Credit Register database does not contain information on inside and outside collateral. Therefore, this paper will only discuss empirical results concerning the role of collateral and guarantees on bank loan interest rates charged to Italian small-sized firms.

This paper uses unbalanced sample data on 197 individual Italian bank and 326,534 firm data drawn from the Central Credit Register at the Bank of Italy for the fourth quarter of the years 2006–2009.

The empirical strategy is as follows.

Firstly, the paper estimates a bivariate probit model to understand how the crisis, together with the borrower observed riskiness and other variables, influenced the probability of observing secured loans. This analysis seems to confirm that guarantees are associated with riskier borrowers.

Secondly, the paper estimates a multilevel model in which the dependent variable, i.e. the spread between the bank loan and overnight interbank interest rates, is regressed on loan-contract, individual-firm and -bank characteristics, and on time dummies. Our results show that collateral helps reduce loan interest rates of Italian producer households. As for personal guarantees, we find that their impact on loan cost is overall negative and statistically significant once we account for the crisis.

The paper is organized as follows. Section 2 reviews the theoretical and empirical literature, while Sect. 3 describes the data used and provides some descriptive statistics; Sect. 4 describes the empirical model, the estimation strategy and discusses the findings. Section 5 concludes.

2 Review of the Literature

The role of collateral and guarantees on bank lending activity has been widely discussed. In this section we focus on the main contributions that analyze the theoretical and empirical relationship between guarantees and loan interest rates.

Theoretically, under perfect information, the bank can distinguish between different types of borrowers, and has perfect knowledge about their investment projects. In this setting guarantees are used to reduce the potential loss that the bank would incur in the case of borrowers' default. According to the "sort by observed risk paradigm" observably risky borrowers are asked to pledge collateral while observably safe borrowers are not (Stiglitz and Weiss, 1981, Berger and Udell 1990).

Under asymmetric information guarantees play a role in solving different problems that may arise at loan origination (hidden information) or after the loan is granted (hidden action).

A hidden information-adverse selection problem arises in situations in which banks cannot discern the *ex-ante* riskiness of the entrepreneur. Without guarantees, the average loan rate would be higher than the optimal rate for safe borrowers, and only riskier borrowers would apply for banks loans. In these situations, guarantees act as a signaling device to distinguish the *ex-ante* riskiness of the entrepreneur: the low risk borrower will choose a contract with guarantees in order to take advantage of a lower interest rate (Stiglitz and Weiss 1981; Bester 1985, 1987; Besanko and Thakor 1987).

A hidden action-moral hazard problem arises when banks cannot observe the borrower behavior after the loan has been granted. In these situations, guarantees are used as an incentive device, and reduce the debtor incentive to strategically default. As Boot et al. (1991) have shown, if there is substitutability between the borrower quality and action, the riskier borrower pledges more guarantees, while the safe borrower gets an unsecured loan. Guarantees are used as an incentive device: secured loans are made to riskier borrowers to reduce the debtor incentive to strategically default.

Therefore, three alternatives define the relation between loan interest rate and guarantees.

First, the risk premium and guarantees are negative related if guarantees are used as a signaling device to solve the adverse selection problem.

Second, the relation between guarantees and risk premium is still negative if guarantees are used as an incentive device to reduce moral hazard, and guarantees more than compensate the borrower risk.

Finally, the relation between guarantees and risk premium is positive if guarantees are used as an incentive device to reduce moral hazard, and guarantees do not compensate the borrower risk.

Empirical results on the impact of collateral and personal guarantees on the loan rate are mixed.

There are studies that find a positive impact of guarantees on interest rates.

Berger and Udell (1990, 1995), by using a large dataset on American commercial loans, have found that interest rates on secured loans are on average higher than those on unsecured loans. Guarantees are most often associated with riskier borrowers, riskier loans, and riskier banks. This result supports the idea that observably riskier borrowers are asked to pledge more guarantees to mitigate the moral hazard problem.

Casolaro et al. (2008), using information on syndicated credit facilities granted to borrowers of over 80 countries between 1990 and 2001, have also found that secured loans have larger interest rate spreads than unsecured loans.

Ono and Uesugi (2009), who have investigated the determinants of collateral and personal guarantees in the small business loan market in Japan, have reached similar results. The authors have found that both collateral and personal guarantees increase interest rates charged on loans and the finding is robust to potential guarantees endogeneity. The authors conclude that guarantees are useful in mitigating the moral hazard as they are more likely to be pledged by riskier borrowers.

Ogawa et al. (2010) use a matched sample of Japanese small firms and main banks, and investigate bank-firm relationships in the early 2000s. The authors have concluded that personal guarantees positively affect the interest rates charged on small firms lending.

Other studies find a negative impact of guarantees on interest rates.

Harhoff and Körting (1998) present a study of the lending relationship between banks and SMEs in Germany. The authors have concluded that interest rates on SME secured loans are lower than those on SME unsecured loans.

Jimènez et al. (2006) have uncovered direct evidence of a negative association between collateral and the borrower's risk.

In a principal-agent setting, John et al. (2003) have concluded that guarantees decrease the riskiness of a given loan, and that collateralized debt has a higher yield than general debt, after controlling for credit rationing.

Pozzolo (2004), Bonaccorsi di Patti (2006) and Calcagnini et al. (2012) provide evidence for Italy.

Pozzolo (2004) has found that collateral are not statistically related to the borrower risk. He interprets this finding as potentially consistent with the hypothesis that inside collateral is used as a signaling device to solve the adverse selection problem. On the other hand, he finds that personal guarantees are more likely to be asked for when the borrower is *ex-ante* riskier. However, once the borrower's riskiness is controlled for, both collateral and personal guarantees reduce the interest rate charged on loans.

Bonaccorsi di Patti (2006) presents a detailed analysis of the widespread use of collateral and personal guarantees in the loan market for Italian firms. The article finds that guarantees are more likely to be pledged by riskier firms, and that personal guarantees and collateral are positively correlated. Moreover, the article finds that internal and external collateral are substitutes.

3 Data and Summary Statistics

The empirical analysis uses information on loans to a large sample of Italian nonfinancial small firms (producer households). Data has been taken from two sources: the Banks' Supervisory Reports to the Bank of Italy (Segnalazioni di Vigilanza) and the Central Credit Register (Centrale dei Rischi). The first source is used for data on banks' balance sheets. The second contains information on single

Table 1 Producer households' loans by type of guarantees (% distribution)

	1999	2000	2001	2002	2003	2004	2005	2006	2007	2008	2009
Collateral	33.7	35.6	36.2	38.2	43.1	46.1	48.6	47.7	51.6	51.5	51.9
Personal guarantees	39.3	38.6	36.3	34.6	30.8	30.2	28.6	29.0	25.1	24.7	23.3
Unsecured	27.0	25.8	27.4	27.2	26.1	23.7	22.9	23.3	23.3	23.7	24.9

Source: Supervision reports – Bank of Italy

bank loans, the interest rates charged and the value of the assets posted as guarantees.[3,4]

We use data on producer households for the period 2006–2009 for a total of 326,534 small-sized firms and 197 banks.

Table 1 shows the distribution (% values) of producer households' loans by type of guarantees for the period 1999–2009. The share of loans granted by collateral (mainly mortgages) constantly increased until 2007 due to the growth of the Italian real estate market. From 2007 this share has been constant as a consequence of the economy slowdown and the decreasing return of real estate investments (Panetta and Signoretti 2010). The faster rate of growth observed in the first type of loans has been compensated by the share of loans granted by personal guarantees, that are potentially riskier than collateral as they represent a generic claim on the wealth of the grantor, who has therefore a large degree of freedom and could possibly default

[3] The Central Credit Register is regulated by the resolution adopted by the Credit Committee on 29 March 1994 pursuant to Articles 53, 67 and 107 of the Banking Law. The following participate in this centralized service:

– Banks entered in the register referred to in Article 13 of the Banking Law;

– Financial intermediaries entered in the register of banking groups and/or the special register referred to in, respectively, Articles 64 and 107 of the Banking Law that engage exclusively or primarily in financing activity. Financial intermediaries more than 50% of whose financing activity consists of consumer credit are exempted. Consequently, the group of financial intermediaries reporting to the Central Credit Register is not identical to the group that transmits supervisory returns.

Participating intermediaries also report the exposures of foreign branches to borrowers resident in Italy. All the statistical distributions take such loans into account.

Once a month intermediaries are required to report each customer's debtor position, comprising both individual and joint liabilities (joint accounts and partnerships).

The whole position relative to a given customer must be reported where even one of the following conditions applies: the sum of credit granted or used for all loans and guarantees granted to the customer is at least 30,000 euros; the total value of personal guarantees provided by the customer is at least 30,000 euros; the customer's position is classified among bad debts or is written off during the reference month, regardless of the amount; the face value of factoring claims the intermediary has acquired from the customer is at least 30,000 euros; the value of the transactions carried out by the intermediary on behalf of third parties is at least 30,000 euros.

Where a report is made because one of the above conditions applies, it must cover all the outstanding positions of the customer in question (Bank of Italy 2010, p. 117).

[4] Before 2009 the threshold was 75,000 euros. Once we account for this change, the findings shown in Tables 2, 3, and 4 are actually reinforced.

Table 2 Number of loans and average loan size

	Total loans			
	Number of loans	Change (%)	Average loan size	Change (%)
2006	245,028		152,287.8	
2007	261,746	6.82	157,617.8	3.50
2008	253,404	−3.19	165,106.7	4.75
2009	157,543	−37.83	167,236.6	1.29

Source: Our calculations on Bank of Italy data

Table 3 Number of loans and average loan size by type of guarantee

	Loans secured by personal guarantees				Loans secured by collateral			
	Number	Change (%)	Average loan size	Change (%)	Number	Change (%)	Average loan size	Change (%)
2006	108,835		154,498.6		166,746		169,630.4	
2007	114,417	5.13	160,444.9	3.85	180,308	8.13	175,248.8	3.31
2008	116,830	2.11	168,722.1	5.16	171,418	−4.93	184,409.7	5.23
2009	93,166	−20.26	165,636.1	−1.83	95,102	−44.52	197,473.7	7.08

Source: Our calculations on Bank of Italy data

on it. Finally, all other types of unsecured loans (labelled as 'Unsecured' in Table 1) have been relatively constant, especially during the most recent years.

As expected, the financial crisis has negatively affected the number of loans. In our database, the latter showed a 37.83% decrease from 2008 to 2009, while the average loan size increased by 1.27% (see Table 2).[5] If we distinguish between loans secured by collateral and by personal guarantees, descriptive statistics show (a) a decrease in the number and average size of loans granted by personal guarantees (−20.26% and −1.83%, respectively); (b) a large drop in the number of loans granted by collateral (−44.52%), but an increase in their average size (+7.08%) (see Table 3). This dynamic can be mainly explained by negative demand effects (fewer investments, fewer mortgages and fewer collateralized loans), but also by banks tightening their credit, and higher bank risk aversion.

During the crisis, following the official interest rate trends, the cost of credit decreased significantly and in 2009 interest rates were 3.9% points lower than the 2008 levels for variable interest rate contracts, and 3.1% points lower than fixed interest rate contracts (Panetta and Signoretti 2010). As for producer households, interest rates on secured loans decreased more than interest rate on unsecured loans (see Table 4). However, interest rate spreads charged to producer households have

[5] The large decrease in the number of contracts in the database could be partially due to statistical changes which occurred in personal or firm data, and industry reclassifications which occurred in 2009. However, variables' trend in our database are consistent with those observed in the data referred to the whole population of banks and households businesses. Indeed, from 2008 to 2009, the total number of lending contracts in Italy decreased by 20.47%. See http://bip.bancaditalia.it/4972unix/homebipita.htm

Table 4 Interest rates and yearly changes by type of guarantee

	Loans secured by personal guarantees		Loans secured by collateral		Unsecured loans	
	Average interest rate	Change	Average interest rate	Change	Average interest rate	Change
2006	6.00		5.32		6.70	
2007	6.80	0.80	6.22	0.90	7.33	0.63
2008	7.00	0.20	6.42	0.20	7.46	0.13
2009	4.72	−2.28	4.10	−2.32	5.47	−1.98

Source: Our calculations on Bank of Italy data

Table 5 Interest rate spreads and yearly changes by type of guarantee

	Total loans		Loans secured by personal guarantees		Loans secured by collateral		Unsecured loans	
	Interest rate spreads	Change	Interest rate spreads	Change	Interest rate spreads	Change	Interest rate spreads	Change
2006	2.41		2.62		1.98		3.21	
2007	2.62	0.21	2.82	0.20	2.28	0.30	3.26	0.05
2008	3.68	1.06	3.89	1.07	3.34	1.06	4.27	1.01
2009	4.48	0.80	4.52	0.63	3.94	0.60	5.27	1.00

Source: Our calculations on Bank of Italy data

been increasing since 2006, and they have been higher for unsecured loans than for loans secured by guarantees, especially for collateralized loans (see Table 5).

Summing up, descriptive statistics show a correlation between loan interest rates and collateral and personal guarantees, as secured loans demonstrate both lower interest rates and lower increases in the spreads during the observed period.

To disentangle the direct effect of collateral and personal guarantees (and their interaction with the crisis) on the cost of credit, we estimate multilevel model relating interest rate spreads to loan contracts, firm, and bank characteristics.

4 Model Specification and Estimation Strategy

4.1 Determinants of Personal Collateral and Guarantees: Bivariate Probit Model

The empirical strategy consists of a two-step analysis. In the first step, we aim at estimating the determinants of collateral and personal guarantees. In particular, we want to verify if previous findings about the role of collateral and personal guarantees in solving adverse selection and moral hazard problems are confirmed, i.e. high quality debtors may use collateral as a signal device, while riskier borrowers are requested to post personal guarantees as an incentive device to solve moral hazard problems.

Table 6 Summary statistics of the sample variables

Variable	Observations	Mean	Median	SD	Min.	Max.
SPD	917,721	3.17	2.81	1.71	0	12.12
RISK	917,721	0.03	0	0.17	0	1
COLL	917,721	0.59	0.81	0.45	0	1
PERS	917,721	0.46	0	0.54	0	2
DOUBLEG	917,721	0.29	0	0.45	0	1
LOAN_S	917,721	0.91	0.70	0.73	0.04	5.79
FIRM_S	917,721	0	0	0.06	0	1
LARGE	917,721	0.13	0	0.34	0	1
LEND_REL	917,721	0.57	1	0.49	0	1
NUM_REL	917,721	1.42	1	0.89	1	20

Source: Our calculations on Bank of Italy data

We assume that the conditional probability of the firm to post guarantees, Pr(*GUAR* = *1*/*X*), given a cumulative distribution function Φ (.), depends on loan, firm and bank characteristics. Moreover, we add time dummies to capture the impact of the financial crisis.

Collateral and personal guarantees are jointly determined and depend on the same set of variables. Therefore, the error correlation between the two types of guarantees may be different from zero, and we estimate the following bivariate probit model:

$$p_{ij} = \Pr\left(GUAR_{ij,t} = 1|X\right) = \Phi(X'\beta)$$

with

$$\begin{aligned} X'\beta = \beta_0 &+ \beta_1 RISK_{ijt} + \beta_2 LOAN\_S_{ijt} + \beta_3 FIRM\_S_{it} + \beta_4 LEND\_REL_{ijt} \\ &+ \beta_5 NUM\_REL_{it} + \beta_6 BANK\_S_j + \beta_7 CENTRAL_i + \beta_8 SOUTH_i \\ &+ \beta' TIME\_D_t + \varepsilon_{ijt} \end{aligned} \quad (1)$$

Variables used in model (1) are described in detail in the Data Appendix.

Table 6 reports data summary statistics.

Columns (1) and (2) of Table 7 report the marginal effects of the bivariate probit model (1), in which the likelihood ratio test rejects the null of zero correlation between the errors of the two probit models (see Table 7 *rho* labeled LR test).

Results support the hypothesis that guarantees are associated to riskier borrowers. We capture customer risk by using a measure of observed risk (RISK) given by the "substandard" loan of the firm in temporary difficulties.[6]

[6] In empirical literature the interest rate charged on the loan is often used as the measure of customer riskiness. The use of an observed measure of customer riskiness such as RISK allows controlling for the endogeneity issue concerning guarantees.

Table 7 The determinants of personal collateral and guarantees. Bivariate probit marginal effects

VARIABLES	(1) Personal guarantees	(2) Collateral	(3) Personal guarantees and large	(4) Collateral and large
RISK	0.014***	0.054***	0.014***	0.050***
	(0.003)	(0.003)	(0.003)	(0.003)
LOAN_S	−0.002***	0.252***	−0.012***	0.309***
	(0.001)	(0.001)	(0.001)	(0.001)
FIRM_S	−0.001	−0.458***		
	(0.008)	(0.009)		
LEND_REL	0.059***	0.038***	0.060***	0.030***
	(0.001)	(0.001)	(0.001)	(0.001)
NUM_REL	−0.004***	−0.153***	−0.003***	−0.154***
	(0.001)	(0.001)	(0.001)	(0.001)
CENTRAL	0.019***	0.081***	0.021***	0.072***
	(0.001)	(0.001)	(0.001)	(0.001)
SOUTH	−0.127***	0.188***	−0.124***	0.174***
	(0.001)	0.001	(0.001)	0.001
BANK_S	−0.027***	−0.001	−0.026***	−0.008***
	(0.001)	(0.001)	(0.001)	(0.001)
2007	−0.028***	0.068***	−0.027***	0.056***
	(0.001)	(0.001)	(0.001)	(0.001)
2008	−0.010***	0.061***	−0.009***	0.052***
	(0.001)	(0.001)	(0.001)	(0.001)
2009	0.102***	0.030***	0.102***	0.022***
	(0.001)	(0.001)	(0.001)	(0.001)
LARGE			0.030***	−0.216***
			(0.002)	(0.002)
Observations	917,721		917,721	
LR test rho = 0	0.00 (p-value reported)		0.00 (p-value reported)	
AIC	2,234,231		2,227,902	
BIC	2,234,501		2,228,171	

Standard errors in parentheses. *** $p < 0.01$, ** $p < 0.05$, * $p < 0.1$

The estimated coefficient of RISK is positive and statistically significant in both columns: riskier borrowers have a higher probability of posting collateral or personal guarantees.

While the impact of loan size (LOAN_S) on personal guarantees is negative but small in absolute values (column (1)), LOAN_S increases the probability of loans to be secured by collateral (column (2)): the result is likely to be driven by the presence of mortgages which, by the Italian code, have to be collateralized.[7]

The stronger bargaining power of large firms (FIRM_S, proxied by the loan size) with respect to firms of smaller size is captured by the negative marginal effect on the probability of posting collateral.

[7] In our data we do not have enough information to distinguish between different types of loans.

Long-term lending relationships (LEND_REL) between banks and customers negatively affect the probability of posting both personal guarantees and collateral. On the one hand, a long-term banking relationship may benefit the borrowers by helping to build trust between borrowers and lenders, and consequently to reduce moral hazard. If guarantees are asked to solve moral hazard problems, the probability of posting guarantees decreases the longer the lending relationship (Boot and Thakor 1994). On the other hand, longer lending relationships could be associated with a higher use of collateral if long-term relationships generate more severe hold-up problems (Ogawa et al. 2010). The findings suggest that the negative effects of the hold-up problem dominate the benefits of the relationship lending.

Furthermore, increasing multiple lending relationships (NUM_REL) negatively affect the probability of posting both collateral and personal guarantees. The result is consistent both with the hypothesis that "banks are unwilling to require a guarantee on their loans if this has the side effect of making implicitly available to competing lenders the result of their screening activity" (Pozzolo 2004, p. 14) and with the fact that firms take actions against the monopoly power of a main bank, and eventually get better contract conditions, in the presence of multiple-bank relationships (Ogawa et al. 2010).

Customer regional location plays a significant role on the likelihood of posting both personal guarantees and collateral. Compared to customers located in the North of Italy, loans provided by in Central Italy (CENTRAL) show a higher probability of posting guarantees or collateral. Located in the South (SOUTH) show a positive marginal effect on collateral and a negative marginal effect on personal guarantees: this result is likely due to the riskiness of this type of guarantees in regions affected, among others, by high criminal rates.

The probability for firms to post personal guarantees is lower if banks are large (BANK_S $= 1$). The finding is consistent with the theoretical and empirical literature according to which larger banks have a higher ability to evaluate customer risk and, therefore, screen riskier investment projects. Particularly, secured loans may be considered by the lender as an alternative to screen and evaluate borrower or loan riskiness. Smaller banks generally have a lower level of expertise and scarce resources to evaluate the economic loan risks. Therefore, they have more incentives to use collateral instead of undertaking a project evaluation (Manove and Padilla 1999; Manove et al. 2001; Jiménez et al. 2006).

The economic and financial crisis increases the probability of loans to be secured, especially during the year 2009.

To capture the impact of both large- and medium-sized firms on interest spreads, columns (3) and (4) of Table 7 show the estimated coefficients of model (1) in which we use an alternative measure of firm size with respect to FIRM_S. LARGE is again a dummy variable that takes value equal to one if the loan granted is greater or equal to 250,000 euro. Producer households are normally of small size. In the sample, 13.55% of loans are greater or equal to the 250,000 euro threshold while only less than 0.5% are greater or equal to 1,000,000 euro. The estimates confirm previous findings that larger firms have a lower probability to post collateral. In this case, however, the probability to post personal guarantees increases.

4.2 Guarantees and Loan Interest Rates: Multilevel Models

The previous Section analyzed the determinants of collateral and personal guarantees, and results show that riskier borrowers have a higher probability of posting both collateral and personal guarantees as an incentive device to solve moral hazard problems.

This Section focuses on the impact of guarantees on loan interest rates.

Here, we assume that the bank loan interest rate is a positive function of:

$$r_{ij} = \mathrm{F}(r(L),\ markup_{ij},\ riskpremium_{ij} \tag{2}$$

where r_{ij} is the interest rate charged to customer i by bank j,$r(L)$ is the market interest rate, $markup_{ij}$ is the mark-up over the market interest rate, and $riskpremium_{ij}$ captures loan and customer riskiness.

Here, $r(L)$ is the overnight interest rate which is the same for all banks.

The mark-up term captures banks' market power. We assume that the mark-up varies across customers inversely with the customer size (FIRM_S) and directly with the bank size (BANK_S). Moreover, the number of bank relationships (NUM_REL) and the time length (number of years) of the lending relationship (LEND_REL) may also affect the mark-up. While we expect a negative impact for NUM_REL, the impact of the length of the lending relationship is not known *a priori*. On one side, a longer lending relationship should increase loan interest rates by generating an information monopoly that enables banks to extract rents from borrowers. On the other, a long-term banking relationship may benefit the borrowers: borrowers pay higher interest rates and pledge guarantees early in the relationship, but, once their first project is successful, they are awarded with unsecured loans and lower loan rates (Boot and Thakor 1994).

The risk premium is the interest rate component that positively depends on customer and loan riskiness. We capture customer risk by means of a measure of observed risk, RISK, as described in Sect. 4.1.

Loan riskiness depends positively on loan size (LOAN_S). It is possible that LOAN_S may capture firm size. If this were the case the estimated coefficient of this variable may have negative sign.

Moreover, we make use of additional information on the presence of guarantees to control for customers' risk. Specifically, we use the relative (to the loan size) amount of collateral (COLL) and personal guarantees (PERS) posted, and a dummy (DOUBLEG) to capture the contemporaneous presence of both types of guarantees. However, as described in Sect. 2, the impact of guarantees on interest rate is not defined *a priori*. Indeed, guarantees may be used as a signal of high quality debtor, and therefore we should expect a negative impact on interest rate; or riskier borrowers may post guarantees, and therefore we should expect a positive impact of guarantees on interest rate.

Finally, we control for customers' geographical location by means of three dummy variables (NORTH, CENTRAL and SOUTH).[8]

Since $r(L)$ and r_{ij} are highly correlated, we opted for a slightly different version of model (2) where the dependent variable is the spread between the two interest rates ($SPD_{i,j}, = r_{ij} - r(L)$).

As explanatory variables we have three main groups of variates:

- A vector $X_{i,j,t}$ containing the characteristics of each loan contract: COLL, PERS, DOUBLEG and LOAN_S;
- A vector $F_{i,t}$ containing firm characteristics: RISK, NUM_REL, LEND_REL, FIRM_S, CENTRAL and SOUTH
- A vector $B_{j,t}$ containing bank characteristics: BANK_S.

Furthermore, our model includes time-dummy variables to identify the impact of the economic crisis and two interaction variables that are expected to capture the impact of the financial crisis on collateral and personal guarantee requirements CRISIS*COLL and CRISIS*PERS, respectively.

The empirical equation takes the following form:

$$\begin{aligned} \text{SPD}_{i,j,t} = \; & \beta_0 + \beta_1' X_{i,j,t} + \beta_2' F_{i,t} + \beta_3' B_{j,t} + \beta_4' TIME\_D_t \\ & + \beta_5 CRISIS^* COLL_{i,j,t} + \beta_6 CRISIS^* PERS_{i,j,t} + u_{i,j,t} \end{aligned} \tag{3}$$

The subscript i refers to firms, j to banks, t to time periods. $u_{i,j,t}$ is a disturbance with a multiway error-components structure:

$$u_{i,j,t} = \alpha_i + \lambda_{i,j} + \varepsilon_{i,j,t}$$

where α_i, $\lambda_{i,j}$ and $\varepsilon_{i,j,t}$ are assumed to be i.i.d., and are mutually independent.

Antweiler (2001) derived the maximum likelihood estimator for panel data with unbalanced hierarchies. We deal with a single-nested panel in which firms may be grouped by banks, and estimate a "mixed effects" model in which a fixed-effects approach is used to estimate regression coefficients and a random-effects approach is used for the low-level group, i.e. banks.

[8] To determine if our interest rate model can be correctly identified as a supply function, we should assume that the variance of the stochastic term in the loan offer function is smaller than the corresponding variance in the loan asking function. This assumption seems acceptable given that, for instance, a given bank forces its lending officers to follow certain common techniques of credit analysis that may result in more precision in processing lending application (i.e. in a lower stochastic variance). Diversely, borrowers are subjected to industry specific seasonal and cyclical shocks; moreover, firm treasurers are not compelled to behave similarly when they apply for loans. Both reasons imply a larger stochastic variance for the loan asking function (see Hester (1967), p. 132).

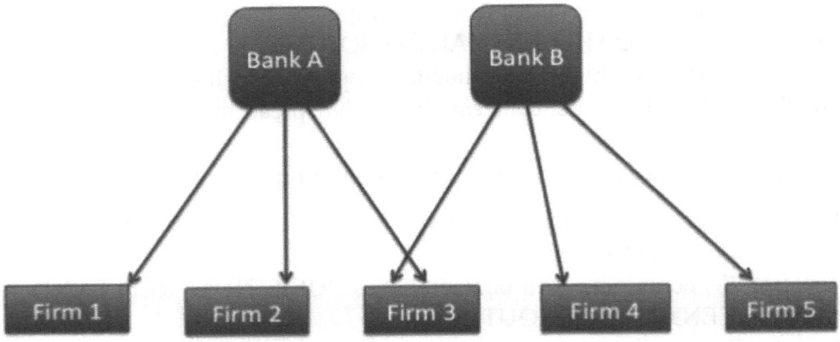

Fig. 1 Nested panel data model

The multilevel analysis assumes that the latent variables, or random effects, can be interpreted as unobserved heterogeneity at the different levels inducing dependence among all lower-level units in the same higher-level unit. Whereas random intercepts represent heterogeneity between clusters in the overall response, random coefficients represent heterogeneity in the relationship between the response and explanatory variables (Rabe-Hesketh et al. 2004). Figure 1 represents the data structure. We have information on bank loans granted to firms. Each bank grants loans to many firms. The paper estimates model (3) parameters both under the assumption that the intercept is random and the overall response varies across banks; and under the assumption that the slope is also random, i.e. the impact of guarantees and collateral on the loan interest rate varies across banks. Moreover, we allow for correlation among random effects by assuming an unstructured variance-covariance matrix.

In the dataset some firms may have multiple bank relationships in each year (Firm 3 in Fig. 1).[9]

Table 8 shows the estimated coefficients of the multilevel models. The likelihood ratio test rejects the null of no random effects for all specifications (see Table 8 *re* labeled LR test, p-value reported).

Different model specifications are shown in Table 8. First, we estimated model (3) both under the assumption that the intercept is random, and under the assumption that the slope is also random, i.e. the impact of guarantees and collateral on the loan interest rate varies across banks. Testing the two specifications by means of the Akaike Information Criterion (AIC) and the Bayesian Information Criterion (BIC), according to which "small is better", the latter specification is preferred to the former. Therefore, column (1) of Table 8 shows the estimated coefficients of the

[9] The panel data is not a pure nested model, as we have some firms that have loans from different banks in each period. Therefore, for robustness checks purposes, we estimate model (3) considering only firms that do not have multiple bank relationships (NUM_REL = 1) in each year. Estimates confirm the findings of Tables 8.

Table 8 The determinants of bank loan interest rates (SPD). Multilevel models

	(1) Baseline model	(2) Guarantees and crisis interaction	(3) Guarantees and risk interaction	(4) Alternative firm size "Large"	(5) IV multilevel model
RISK	0.661***	0.659***	1.273***		0.624***
	(0.008)	(0.008)	(0.016)		(0.009)
COLL	−1.182***	−1.208***	−1.186***	−1.213***	
	(0.045)	(0.045)	(0.046)	(0.045)	
PERS	−0.018*	0.013	0.017	0.010	
	(0.010)	(0.010)	(0.010)	(0.010)	
DOUBLE_G	0.172***	0.167***	0.167***	0.170***	0.061***
	(0.005)	(0.005)	(0.005)	(0.005)	(0.004)
LOAN_S	−0.150***	−0.149***	−0.151***	−0.111***	−0.156***
	(0.002)	(0.002)	(0.002)	(0.003)	(0.002)
LEND_REL	0.063***	0.065***	0.066***	0.065***	0.071***
	(0.003)	(0.003)	(0.003)	(0.003)	(0.003)
NUM_REL	0.058***	0.057***	0.059***	0.058***	0.073***
	(0.002)	(0.002)	(0.002)	(0.002)	(0.002)
FIRM_S	0.053**	0.050**	0.054**		0.062**
	(0.022)	(0.022)	(0.022)		(0.026)
BANK_S	−0.012	−0.008	−0.009	−0.007	0.043
	(0.056)	(0.056)	(0.056)	(0.056)	(0.106)
CENTRAL	0.053***	0.052***	0.051***	0.051***	0.044***
	(0.006)	(0.006)	(0.006)	(0.006)	(0.006)
SOUTH	0.130***	0.132***	0.129***	0.129***	0.102***
	(0.004)	(0.004)	(0.004)	(0.004)	(0.005)
2007	0.231***	0.232***	0.232***	0.233***	0.233***
	(0.004)	(0.004)	(0.004)	(0.004)	(0.004)
2008	1.257***	1.247***	1.240***	1.250***	1.259***
	(0.004)	(0.006)	(0.006)	(0.006)	(0.004)
2009	1.867***	1.865***	1.856***	1.867***	1.878***
	(0.004)	(0.006)	(0.006)	(0.006)	(0.005)

(continued)

Table 8 (continued)

	(1) Baseline model	(2) Guarantees and crisis interaction	(3) Guarantees and risk interaction	(4) Alternative firm size "Large"	(5) IV multilevel model
CRISIS*COLL		0.061***	0.073***	0.063***	
		(0.006)	(0.006)	(0.006)	
CRISIS*PERS		−0.060***	−0.057***	−0.060***	
		(0.005)	(0.005)	(0.005)	
RISK*COLL			−0.873***		
			(0.019)		
RISK*PERS			−0.104***		
			(0.014)		
LARGE				−0.105***	
				(0.006)	
IVCOLL					−0.962***
					(0.077)
IVPERS					0.008
					(0.021)
Constant	2.960***	2.961***	2.947***	2.942***	2.857***
	(0.047)	(0.047)	(0.047)	(0.047)	(0.063)
Observations	917,721	917,721	917,721	917,721	917,721
LR test $re = 0$	0.00	0.00	0.00	0.00	0.00
AIC	3,107,039	3,106,808	3,104,643	3,106,526	3,194,662
BIC	3,107,297	3,107,090	3,104,948	3,106,807	3,194,920
Num. of banks	197	197	197	197	197

Standard errors in parentheses *** $p < 0.01$, ** $p < 0.05$, * $p < 0.1$

multilevel model (3) with random intercept and random slope, but without the interaction variables CRISIS*PERS and CRISIS*COLL.

Results show that the presence of collateral (COLL) decreases the interest rate, while the presence of personal guarantees (PERS) is statistically significant only at 10% level. These findings are consistent with the idea that once banks control for the borrower risk, collateral decreases interest rates. Moreover, producer households, being of small size, suffer from informational opacity more than larger firms, and collateral might act as a positive signalling device (Berger and Udell 1998). However, the estimated coefficient of DOUBLEG is positive and statistically significant: it is likely that riskier borrowers are requested to post both types of guarantees.

Loan size (LOAN_S) and interest rates show an inverse relationship. In this case the scale effect of the loan size more than counterbalances the potentially higher risk associated to loans of larger size.

As for firm characteristics, the longer the lending relationship (LEND_REL) is, the higher the interest rate. The finding is not new to the empirical literature (see Harhoff et al. 1998; Petersen and Rajan 1994 for a survey of the empirical literature). Chakravarty and Yilmazer (2009) assert that the overall granting process is a sequential process which unfolds in three stages: application, decision and rate setting. The authors find that the lending relationship matters only in the first and second stages, i.e.: conditional on being approved, relationships are not important in determining the loan rate. Similarly, Petersen and Rajan (1994) do not find statistical evidence that the strength of the lender-borrower relationship is correlated with cheaper credit. Moreover, our dataset contains different types of loans for which reputation and relationship effects may be less important (Berger and Udell 1995). Therefore, the length of the lending relationship may capture not only the strength of the bank-borrower relationship, but also a monopoly power of a main bank that asks for higher interest rates.

The ex-ante (or observed) firm riskiness is captured by the variable RISK whose estimated coefficient, as expected, is positive and statistically significant. Furthermore, the number of lending relationships (NUM_REL), which may be also interpreted as a measure of borrower riskiness, increases interest rates.

As for firm size, our results show that the former has a positive influence on interest rates, but this result may be due to the way the FIRM_S variable was constructed. Indeed, FIRM_S is a binary dummy variable which takes a value of 1 when the loan value is equal or greater €1,000,000 and 0 when the loan value is less than €1,000,000.[10] Producer households are normally of small size; therefore this variable might capture the higher risk associated to larger loans. Furthermore, firm size may be more efficiently captured by the LOAN_S.

Customers located in the Centre and the South of Italy are charged a higher interest rate than customers in the North. Finally, time dummies account for the

[10] This threshold is used in the statistics of European Central Bank and in several Bank of Italy papers.

effects of the financial crisis on interest rates. Their coefficients are highly statistically significant and positive reflecting the increase in interest rates between 2006 and 2009.

To account for the potential interaction of the crisis with the presence of collateral and guarantees, column (2) of Table 8 shows the estimated coefficients of model (3).

Overall, estimates shown in column (2) confirm the findings of column (1). Specifically, the estimates show that loans secured by personal guarantees pay a lower interest rate only during the crisis, as the coefficients of PERS is not statistically significant and the coefficient of CRISIS*PERS is negative and statistically significant at 1% level. The result, together with the negative estimated coefficients of collateral (COLL) show that collateral and guarantees contribute both to reducing loan interest rates and avoiding credit rationing (as shown by descriptive statistics in Tables 3 and 4).

All model specifications find that riskier firms pay higher interest rates. Column (3) shows the estimated coefficients of model (3) in which we further control for the impact of the interaction of firm risk with guarantees on loan interest rates by adding two interaction variables RISK*COLL and RISK*PERS. Estimates show that the negative impact of guarantees on loan interest rates is higher the riskier the firm: *ceteris paribus*, guarantees are a more powerful instrument for *ex-ante* riskier borrowers than for safer borrowers. Indeed, riskier borrowers obtain significantly lower interest rates on secured loans than the interest rate they would be charged on unsecured loans.

To capture the impact of both large- and medium-sized firms on the interest spreads, column (4) shows the estimated coefficients of model (3) in which the variable LARGE is used instead of the variable FIRM_S to measure firm size. The estimated coefficient shows that medium- and large-sized firms pay lower interest rates than smaller firms. Overall, the estimates confirm findings in columns (1)–(3) of Table 8.

To check the robustness of our results column (5) in Table 8 shows the estimated coefficients of the baseline model specification of column (1) when treating collateral and personal guarantees as endogenous variables. Guarantees might be endogenous due to unmeasured and unmeasurable influences acting at firm or bank levels such as the fact that interest rates and guarantees are simultaneously set at the time of a loan approval. [11] In multilevel analysis, the unmeasured influences of omitted variables or measurement error in the fixed part gets incorporated in the random part of the model, thereby violating the assumption of the independence of regressors and model disturbances (Ebbes et al. 2004). If this is the case, the OLS estimated coefficients are biased, and instrumental variables (IV) estimation techniques can be used to reduce the bias (Spencer and Fielding 2000).

[11] All bank level variables in model (3) are exogenous. Therefore, we rule out bank-level endogeneity and only take into account firm level endogeneity.

Table 9 Random-effects parameters

SPD	Coef.	Std. Err.	z	P > z	[95% Conf. Interval]	
Standard dev (COLL)	0.613	0.034	18.18	0.000	0.547	0.679
Standard dev (PERS)	0.116	0.008	15.01	0.000	0.101	0.132
Standard dev (CONSTANT)	0.633	0.034	18.44	0.000	0.566	0.700
Covariance (COLL, PERS)	0.965	0.092	10.52	0.000	0.785	1.145
Covariance (COLL, CONSTANT)	0.173	0.014	12.53	0.000	0.146	0.201
Covariance (PERS, CONSTANT)	1.007	0.096	10.45	0.000	0.818	1.196
Standard dev (RESIDUAL)	1.313	0.001	1,354.34	0.000	1.312	1.315

First of all, to obtain IV estimates of model (3), we used Lewbel's approach to construct 'internal instruments' (Lewbel 1997) and then estimated separate multi-level models for COLL and PERS as function of the instruments and all the other exogenous variables included in the model.

Secondly, we used the predictions of the endogenous variables, IVCOLL and IVPERS, as variables in model (3) to obtain consistent estimates of the fixed-effects parameters. Then, we corrected the coefficients' variance-covariance matrix by means of the 'real' mean square error (Baltagi 2002).[12]

IV estimates confirm the previous finding: collateral decreases interest rates, while personal guarantees have no statistically significant effect on interest rates. This finding is consistent with the fact that personal guarantees are external to the firm and are typically posted by the proprietorship. In the case of producer households, the owners are requested by law to post their personal wealth in case of default and, therefore, the result that the estimated coefficient of personal guarantees is not statistically significant is not surprising.[13] Moreover, even if personal guarantees are external (and therefore should be more powerful in solving adverse selection or moral hazard problems), they are potentially riskier than collateral as they represent a generic claim on the wealth of the grantor, who has therefore a large degree of freedom and could possibly default on it.

Table 8 shows the fixed-effects estimates of model (3). As for guarantees, fixed effects refer to the overall expected effect of the presence of real or personal guarantees on interest rates. However, according to our model specification, random effects also depend on guarantees. Random effects measure whether the impact of guarantees on interest rates differs among banks/from bank to bank.

Table 9 shows the estimated random-effects parameters of model specification shown in column (1) of Table 8. The variance component of the random intercept (Standard dev (CONSTANT)) is statistically significant, meaning that interest rates significantly differ across banks. Furthermore, the slopes of collateral and personal guarantees also show statically significant variance components, (Standard dev

[12] The latter is constructed by taking into account the best linear unbiased prediction of the random part of the baseline model.

[13] The result is consistent with previous studies according to which partnerships have a lower probability of posting personal guarantees than limited liabilities firms (Bonaccorsi di Patti 2006).

(COLL)) and (Standard dev(PERS)), respectively. Therefore, real guarantees and personal guarantees affect loan interest rates through random effects that differ across banks.

All variance components can be used to partition the variance across levels and compute the intra-class correlation coefficient (ICC) of interest rates within a cluster (bank).

Specifically, the intra-class correlation coefficient is the proportion of the interest rate total variance that is attributed to the bank level, and it is equal to:

$$ICC = \frac{0.633^2 + 0.116^2 + 0.613^2}{0.633^2 + 0.116^2 + 0.613^2 + 1.313^2} = 0.31,$$

meaning that the 31% of the interest rate total variance is attributable to bank-level differences (Albright and Marinova 2010).

Finally, the correlations between bank intercepts and slopes (Covariance (COLL, CONSTANT)) and (Covariance (PERS, CONSTANT)) are positive and statistically significant across banks. Therefore, given the negative estimated coefficient of collateral (see Table 8), the decrease in interest rates for a 1% increase in guarantees is larger for banks charging higher average interest rates.

5 Conclusions

This paper analyzed the role of guarantees on loan interest rates charged on producer households' loans. In addition, it tried to understand bank behavior before and during the recent financial crisis. The analysis focused on small businesses since informational opacity is expected to be negatively correlated with firm size. Therefore, small-sized firms are more often affected by either credit rationing and/or higher interest rates than larger firms. These two phenomena typically exacerbate during financial crises. During the period 2008–2009, the number of loans to small businesses decreased by more than 30% as opposed to a 20% decrease in the case of all firms (firms and producer households).

The bivariate probit analysis showed that the probability of loans to be secured increased during the recent financial crises. Moreover, the probability of loans to be secured by collateral or personal guarantees increases with firm riskiness.

Estimates from our interest rate model showed that loan interest rates respond differently to collateral and personal guarantees. Collateral helps reduce loan interest rates charged to small-sized businesses, once we control for borrower and loan riskiness, before and during the financial crisis. As for personal guarantees, our findings showed a positive (a decrease) effect on interest rates only during the financial crisis. Indeed, in normal times, personal guarantees do not give the lender a specific claim on particular assets, and restrict the actions (s)he could take in case of borrowers' bankruptcy. Therefore, they have no effect on loan interest rates.

During troublesome periods, such as the current financial crisis, providing personal guarantees together with collateral is a signal of borrowers' quality that has positive effects on interest rates.

Finally, results showed that – *ceteris paribus* – guarantees are a more powerful instrument for *ex-ante* riskier borrowers than for safer borrowers. Indeed, riskier borrowers obtain significantly lower interest rates on secured loans than interest rate they would be charged on unsecured loans.

Data Appendix

SPD is the spread between the interest rate applied on loan by each bank and the interest rate on overnight interbank deposits. Both interest rates are averages of each year's fourth quarter values.

COLL is the share of each loan guaranteed by collateral. Loans are mainly mortgages granted by banks to the borrower. This variable is a proxy for inside collateral.

PERS is the share of each loan guaranteed by personal guarantees. Personal guarantees are granted by third parties in favour of borrowers. This variable acts as outside collateral.

DOUBLEG is a binary dummy variable that takes a value of 1 when both personal and real guarantees are posted and 0 otherwise.

LOAN_S is the ratio between the amount of loan granted to the firm by each bank in the database and the average size of loan granted to firms of the same sector. It represents a proxy for loan size.

FIRM_S is a binary dummy variable which takes a value of 1 when the amount of loan is equal or greater €1,000,000 and 0 when the value of loan is less than €1,000,000. Alternatively, we use the dummy variable LARGE. The latter is a binary dummy variable which takes a value of 1 when the amount of loan is equal or greater € 250,000 and 0 when the value of loan is less than € 250,000.

RISK is a dummy variable that takes value equal to one if the firm has substandard loans, i.e. the firm is in temporary difficulty. This variable is a measure of *ex ante* (observed) credit risk of the firm.

LEND_REL is a binary dummy variable that takes a value of 1 in the case of a firm-bank relationship three or more years long and a value of 0 in the other cases.

NUM_REL is the number of lending relationships for each firm in each year.

CENTRAL is a binary geographical dummy variable that has a value of 1 for customers with headquarter in Central Italy and 0 otherwise.

SOUTH is a binary geographical dummy variable that has a value of 1 for customers with headquarter in Southern Italy and 0 otherwise.

BANK_S is a binary dummy variable that has a value of 1 for banks which are classified as "major" or "large" according to the classification of Bank of Italy (2008).

CRISIS*COLL represents the interaction between collateral (COLL) and a dummy that is equal to1 in every year of the financial crisis period (2008–2009); it is 0 in the pre-crisis years (2006 and 2007).

CRISIS*PERS represents the interaction between personal guarantees (PERS) and a dummy that is equal to1 in every year of the financial crisis period (2008–2009); it is 0 in the pre-crisis years (2006 and 2007).

References

Albright JJ, Marinova DM (2010) Estimating multilevel models using SPSS, Stata, SAS, and R. Mimeo

Antweiler W (2001) Nested random effects estimation in unbalanced panel data. J Econ 101:295–313

Baltagi BH (2002) Econometrics. Springer, New York

Bank of Italy (2008) Annual report. Rome

Bank of Italy (2010) Statistical bulletin 4/2010

Berger AN, Udell GF (1990) Collateral, loan quality, and bank risk. J Monet Econ 25:21–42

Berger AN, Udell GF (1995) Relationship lending and lines of credit in small firms finance. J Bus 68(3):351–381

Berger AN, Udell GF (1998) The economics of small business finance: the roles of private equity and debt markets in the financial growth cycle. J Bank Finance 22:613–673

Berger AN, Udell GF (2005) Small business and debt finance. In: Zoltan JA, Audretsch DB (eds) Handbook of entrepreneurship research. Springer, New York

Besanko D, Thakor AV (1987) Collateral and rationing: sorting equilibria in monopolistic and competitive credit markets. Int Econ Rev 28(3):671–689

Bester H (1985) Screening vs rationing in credit markets with imperfect information. Am Econ Rev 57:850–855

Bester H (1987) The role of collateral in credit markets with imperfect information. Eur Econ Rev 31:887–899

Bonaccorsi di Patti E (2006) La diffusione delle garanzie reali e personali nel credito alle imprese. Bank of Italy

Boot AWA, Thakor AV (1994) Moral hazard and secured lending in an infinitely repeated credit market game. Int Econ Rev 35(4):899–920

Boot AWA, Thakor AV, Udell GF (1991) Secured lending and default risk: equilibrium analysis, policy implications, and empirical results. Econ J 101(406):458–472

Casolaro L, Focarelli D, Pozzolo AF (2008) The pricing effect of certification on syndicated loans. J Monet Econ 5:335–349

Calcagnini G., Farabullini F., and G. Giombini (2012). The impact of the recent financial crisis on bank loan interest rates and guarantees. MPRA Paper 36682, University Library of Munich, Germany

Chakravarty S, Yilmazer T (2009) A multistage model of loans and the role of relationship. Financ Manage 38(4):781–816

Cowling M (2010) The role of guarantee schemes in alleviating credit rationing in the UK. J Financ Stab 6:36–44

Ebbes P, Bockenholt U, Wedel M (2004) Regressor and random-effects dependencies in multilevel models. Statistica Neerlandica 58:161–178

Harhoff D, Körting T (1998) Lending relationships in Germany: empirical results from survey data. J Bank Finance 22:1317–1353

Hester DD (1967) An empirical examination of a commercial bank loan offer function. In: Hester DD, Tobin J (eds) Studies of portfolio behavior. Wiley, New York, pp 118–170

Jimenez G, Salas V, Saurina J (2006) Determinants of collateral. J Financ Econ 81:255–281

John K, Lynch AW, Puri M (2003) Credit ratings, collateral and loan characteristics: implications for yield. J Bus 76:371–409

Lewbel A (1997) Constructing instruments for regressions with measurement error when no additional data are available, with an application to patents and R&D. Econometrica 65:1201–1213

Manove M, Padilla AJ (1999) Banking (conservatively) with optimists. Rand J Econ 30(2):324–350

Manove M, Padilla AJ, Pagano M (2001) Collateral versus project screening: a model of lazy banks. Rand J Econ 32(4):726–744

Ogawa K, Sterken E, Tokutsu I (2010) Multiple bank relationship and the main bank system. In: Calcagnini G, Saltari E (eds) The economics of imperfect markets, series contribution to economics. Physica/Springer, Berlin/Heidelberg, pp 73–90

Ono A, Uesugi I (2009) The role of collateral and personal guarantees in relationship lending: evidence from Japan's SME loan market. J Money Credit Bank 41(5):935–960

Panetta F, Signoretti FM (2010) Domanda e Offerta di Credito in Italia durante la Crisi Finanziaria. Occasional Papers, 63 Bank of Italy

Petersen MA, Rajan RG (1994) The benefits of lending relationship: evidence from small business data. J Finance 49(1):3–37

Pozzolo AF (2004) The role of guarantees in bank lending. Discussion papers, 528. Bank of Italy

Rabe-Hesketh S, Skrondal A, Pickles A (2004) Generalized multilevel structural equation modeling. Psychometrika 69(2):167–190

Spencer NH, Fielding A (2000) An instrumental variable consistent estimation procedure to overcome the problem of endogenous variables in multilevel models. Multilevel Model News 12(1):4–7

Stiglitz J. E., and Weiss A. (1981). Credit Rationing in Markets with Imperfect Information, American Economic Review, 71(3):393–410.

FSC
www.fsc.org
MIX
Papier aus verantwortungsvollen Quellen
Paper from responsible sources
FSC® C105338

If you have any concerns about our products,
you can contact us on
ProductSafety@springernature.com

In case Publisher is established outside the EU,
the EU authorized representative is:
Springer Nature Customer Service Center GmbH
Europaplatz 3, 69115 Heidelberg, Germany

Printed by Libri Plureos GmbH
in Hamburg, Germany